OCR

Physics

REVISION GUIDE

Stephen Pople
Carol Tear

OXFORD
UNIVERSITY PRESS

Great Clarendon Street, Oxford OX2 6DP

Oxford University Press is a department of the University of Oxford. It furthers the University's objective of excellence in research, scholarship, and education by publishing worldwide in

Oxford New York Auckland Cape Town Dar es Salaam Hong Kong Karachi Kuala Lumpur Madrid Melbourne Mexico City Nairobi New Delhi Shanghai Taipei Toronto

With offices in
Argentina Austria Brazil Chile Czech Republic France Greece Guatamala Hungary Italy Japan South Korea Poland Portugal Singapore Switzerland Thailand Turkey Ukraine Vietnam

Oxford is a registered trade mark of Oxford University Press in the UK and in certain other countries

British Library Cataloguing in Publication Data

Data available

ISBN: 978-0-19913630-8

10 9 8 7 6 5 4 3 2 1

Printed by Bell and Bain Ltd., Glasgow

Acknowledgements:
Authors, editors, co-ordinators and contributors: Stephen Pople, Carol Tear, Claire Gordon.

Project managed by Elektra Media Ltd. Typeset by Wearset Ltd.

Paper used in the production of this book is a natural, recyclable product made from wood grown in sustainable forests. The manufacturing process conforms to the environmental regulations of the country of origin.

CONTENTS

Overview

This e-book is a supplementary resource for use alongside your main course book and to help you with your revision. On its own it is not sufficient for full study of your OCR A Level Physics A course. The e-book is a succinct guide to what you need to know and be able to do and will help you to identify any areas that you need to study more thoroughly.

- You can obtain a copy of the specification and past papers to practise. You can also look at the Practical Skills Handbook from the exam boards' website: www.ocr.org.uk.
- Find out when you will be working on the practical skills and the dates of your exams and plan your revision accordingly.
- Begin revising! The self-assessment questions on pages 72–74 will help you to check your progress.

Specification structure

		OCR Physics A
AS units	Unit 1	**Mechanics** Motion, forces in action, work and energy *1h written exam* *AS 30% A 15%* *60 marks*
	Unit 2	**Electrons, waves, and photons** Electric current, resistance, DC circuits, waves, quantum physics *1h45m written exam* *AS 50% A 25%* *100 marks*
	Unit 3	**Practical skills in physics 1** *Three externally set tasks, internally marked using set mark scheme* *AS 20% A 10%* *40 marks*
A2 units	Unit 4	**The Newtonian World** Newton's laws and momentum, circular motion and oscillations, thermal physics *1h 15min written exam (synoptic questions)* *A 15%* *60 marks*
	Unit 5	**Fields, particles, and frontiers of physics** Electric and magnetic fields, capacitors and exponential decay, nuclear physics, medical imaging, modelling the Universe *2h (synoptic questions)* *A 25%* *100 marks*
	Unit 6	**Practical skills in physics 2** *Three externally set tasks, internally marked using set mark scheme* *A 10%* *40 marks*

Practical assessment

Your practical skills will be assessed at both AS and A2 level.

You are required to carry out 3 tasks at AS, these are:
A qualitative task (10 marks)
A quantitative task (20 marks)
An evaluative task (10 marks)

OCR will provide a selection of each task and your examination centre (school or college) may give you the opportunity to attempt more than one, so that the best mark can be put forward. The tasks are completed with your teacher supervising. Each practical skills unit is marked by your teacher using instructions from OCR and the marking is moderated (checked) by OCR.

The qualitative and quantitative tasks test skills of observation, recording and reaching valid conclusions. The evaluative task tests your ability to analyse and evaluate the procedures followed and/ or the measurements made. You may be asked to suggest improvements to increase the reliability or accuracy of an experiment. This task will be linked to the quantitative task but you will not have to take any more data. (Additional data may be provided.)

Carrying out practical work

GENERAL SKILLS
The general skills you need to practise are
* the accurate reporting of experimental procedures
* presentation of data in tables (possibly using spreadsheets)
* graph drawing (possibly using IT software)
* analysis of graphical and other data
* critical evaluation of experiments

Record
* all your measurements
* any problems you have met
* details of your procedures
* any decisions you have made about apparatus or procedures including those considered and discarded
* relevant things you have read or thoughts you have about the problem.

Define the problem
Write down the aim of your experiment or investigation. Note the variables in the experiment. Define those that you will keep constant and those that will vary.

Suggest a hypothesis
You should be able to suggest the expected outcome of the investigation on the basis of your knowledge and understanding of science. Try to make this as quantitative as you can, justifying your suggestion with equations wherever possible.

Do rough trials
Before commencing an investigation in detail do some rough tests to help you decide on
* suitable apparatus
* suitable procedures
* the range and intervals at which you will take measurements
* consider carefully how you will conduct the experiment in a way that will ensure safety to persons and to equipment.

Remember to consider alternative apparatus and procedures and justify your final decision.

Carry out the experiment
Remember all the skills you have learnt during your course:
* note all readings that you make
* consider carefully the range and intervals at which you make your observations
* take repeats and average whenever possible
* use instruments that provide suitably accurate data
* consider the accuracy to which it is reasonable to quote your observations (how many significant figures are reasonable)
* analyse data as you go along so that you can modify the approach or check doubtful data.

Presentation of data
Tabulate all your observations, remembering to
* include the quantity, any prefix, and the unit for the quantity at the head of each column
* include any derived quantities that are suggested by your hypothesis
* quote measurements and derived data to an accuracy/significant figures consistent with your measuring instruments and techniques. Remember to work out an appropriate unit.
* make sure figures are not ambiguous.

Graph drawing
Remember to
* choose a suitable scale that uses the graph paper fully.
* label your axes with quantity and unit
* use a scale that is easy to use and fills the graph paper effectively
* plot points clearly with a cross using a sharp pencil (you may wish to include 'error bars')
* draw the best line through your plotted points so that the points are scattered evenly about the line or curve
* consider whether the gradient and area under your graph have significance.

Analysing data
This may include
* the calculation of a result
* drawing of a graph
* statistical analysis of data
* analysis of uncertainties in the original readings, derived quantities, and results.

Remember to
* use a large gradient triangle in graph analysis to improve accuracy
* set out your working so that it can be followed easily
* ensure that any quantitative result is quoted to an accuracy that is consisted with your data and analysis methods
* include a unit for any result you obtain.

Evaluation of the investigation
The evaluation should include the following points:
* draw conclusions from the experiment
* identify any systematic errors in the experiment
* comment on your analysis of the uncertainties in the investigation
* review the strengths and weaknesses in the way the experiment was conducted
* suggest alternative approaches that might have improved the experiment in the light of experience.

Use of information technology (IT)
You may have used data capture techniques when making measurements or used IT in your analysis of data. In your analysis you should consider how well this has performed. You might include answers to the following questions.
* What advantages were gained by the use of IT?
* Did the data capture equipment perform better than you could have achieved by a non-IT approach?
* How well has the data analysis software performed in representing your data graphically, for example?

Your Laboratory Notebook
If you write a good report, it should be possible for the reader to repeat what you have done should they wish to check your work.

Use subheadings
These help break up the report and make it more readable. As a guide, the subheadings could be the main sections of the investigation: aims, diagram of apparatus, procedure, etc.

Answering the question

This section contains some examples of types of questions with model answers showing how the marks are obtained. You may like to try the questions and then compare your answers with the model answers given.

MARKS FOR QUALITY OF WRITTEN COMMUNICATION

Quality of written communication is assessed in all units and credit may be restriced if communication is unclear.

You should:

- Make sure that the text is legible and that your spelling, punctuation and grammar are accurate so that the meaning is clear.
- Write your answers in a style that is appropriate for the purpose of answering the question and explaining complex subject matter.
- Organise the information in your answer so that it is clear, and use the proper scientific vocabulary.

---

2 marks if your answer
- uses scientific terms correctly
- is written fluently and/or is well argued
- contains only a few spelling or grammatical errors.

---

1 mark if your answer
- generally uses scientific terms correctly
- generally makes sense but lacks coherence
- contains poor spelling and grammar.

An answer that is scientifically inaccurate, is disjointed, and contains many spelling and grammatical errors loses both these marks.

The message is:
do not let your communication skills let you down.

ALWAYS SHOW YOUR WORKING
It is wise always to show your working. If you make a mistake in processing the data you could still gain the earlier marks for the method you use.

Question 1
Calculation question

The supply in the following circuit has an e.m.f. of 12.0 V and negligible internal resistance.

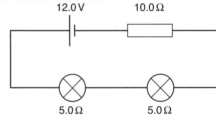

12.0 V   10.0 Ω

5.0 Ω   5.0 Ω

(a) Calculate
   (i) the current through each lamp; *(2 marks)*
   (ii) the power dissipated in each lamp; *(2 marks)*
   (iii) the potential difference across the 10.0 Ω resistor.
                                         (1 mark)

(b) A student wants to produce the same potential difference across the 10.0 Ω resistor using two similar resistors in parallel.
   (i) Sketch the circuit the student uses. *(1 mark)*
   (ii) Determine the value of each of the series resistors used. Show your reasoning. *(3 marks)*

Answer

(a) (i) Current in circuit = e.m.f./total resistance ✓
                 =12.0/20.0
       Current in circuit = 0.60 A ✓
  (ii) Power $= I^2 R$ ✓
             $= 0.60^2 \times 5.0$
      Power $= 1.8$ W ✓
  (iii) p.d. $= IR = 0.60 \times 10.0 = 6.0$ V ✓

(b) (i)

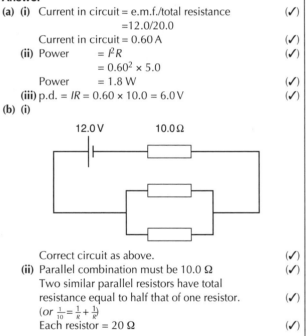

12.0 V   10.0 Ω

Correct circuit as above. ✓

(ii) Parallel combination must be 10.0 Ω ✓
Two similar parallel resistors have total
resistance equal to half that of one resistor. ✓
(or $\frac{1}{10} = \frac{1}{R} + \frac{1}{R}$)
Each resistor = 20 Ω ✓

Question 2

Experiment description

(a) Sketch the apparatus you would use to show different modes of stationary waves in a stretched string or wire. (*2 marks*)

(b) Explain what is meant by 'the fundamental mode' (*2 marks*)

(c) Describe what happens as the frequency of the vibration of the string or wire is increased from zero. (*6 marks*)

Answer

(a)

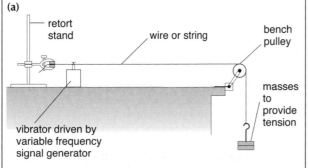

Some means of changing the frequency (✓)

A wire or string in tension (✓)

(b) The fundamental mode is the simplest stationary wave that can be set up. It has the longest wavelength and the lowest frequency. (✓) If the length of the string is *l* the wavelength of the fundamental mode is 2*l*. (✓)

(c) As the frequency is increased to a value *f* which corresponds to the fundamental, a stationary wave is set up with a node at each end of the string, (✓) and one antinode in the centre. (✓) When the frequency is increased to 2*f* the string will vibrate at the 2nd harmonic (✓) and there is a node in the centre as well as at the ends. (✓) The wavelength is now equal to the length of the string. (✓) When the frequency is increased to 3*f* the string vibrates at the 3rd harmonic and this pattern continues as the frequency is increased to higher harmonics (✓).

Question 3

'Show that' question

A length of wire has diameter 0.5 mm and length 50 cm. The resistance is 2.8 Ω. Show that the resistivity of the wire is about 1×10^{-6} Ω m.

Answer

$\rho = RA/l$ (✓) $A = \pi (0.25 \times 10^{-3})^2$ (✓)

$\rho = 2.8 \times 1.96 \times 10^{-7}/0.5$ Ω m

$\rho = 1.1 \times 10^{-6}$ Ω m (❼)

Measurements, uncertainties and graphs

Significant figures

Writing the value of distance $d = 7$ m does not mean the same as writing $d = 7.0$ m.

$d = 7$ m has 1 significant figure, which implies it could be any value between 6.5 m and 7.4 m.

$d = 7.0$ m has 2 significant figures, which implies it could be any value between 6.95 m and 7.04 m.

Examples

1002 kg has 4 significant figures
0.200 g has 3 significant figures
3.07 ml has 3 significant figures
0.012 g has 2 significant figures

Scientific notation

The average distance from the Earth to the Sun is 150 000 000 km.

There are two problems with quoting a measurement in the above form:

- the inconvenience of writing so many noughts,
- uncertainty about which figures are important (i.e. How approximate is the value? How many of the figures are significant?).

These problems are overcome if the distance is written in the form 1.50×10^8 km.

'1.50×10^8' tells you that there are three significant figures – 1, 5, and 0. The last of these is the least significant and, therefore, the most uncertain. The only function of the other zeros in 150 000 000 is to show how big the number is. If the distance were known less accurately, to two significant figures, then it would be written as 1.5×10^8 km.

Numbers written using powers of 10 are in *scientific notation* or *standard form*. This is also used for small numbers. For example, 0.002 can be written as 2×10^{-3}.

Calculations and significant figures

Example

If the distance travelled d is 7.0 m (as above) and the time taken is $t = 2.1$ s (which implies a value in the range $t = 2.05$ s–2.14 s)

The speed $v = \dfrac{d}{t} = \dfrac{7.0\,\text{m}}{2.1\,\text{s}} = 3.3$ m s^{-1} (2sf)

The calculator reads 3.33333… but quoting any more figures would be meaningless because the values of distance and time are not known that accurately. (If you use the range of values for d and t to calculate the biggest and smallest possible values for v, your calculator will show 3,248… and 3.434…)

As a general guide, if the result is to be used in further calculations, it is best to leave any rounding up or down until the end.

Uncertainty

When making any measurement, there is always some *uncertainty* in the reading. As a result, the measured value may differ from the true value. In science, an uncertainty is sometimes called an *error*. However, it is important to remember that it is *not* the same thing as a mistake.

In experiments, there are two types of uncertainty.

Systematic uncertainties These occur because of some inaccuracy in the measuring system or in how it is being used. For example, a timer might run slow, or the zero on an ammeter might not be set correctly.

There are techniques for eliminating some systematic uncertainties. However, this spread will concentrate on dealing with uncertainties of the random kind.

Random uncertainties These can occur because there is a limit to the sensitivity of the measuring instrument or to how accurately you can read it. For example, the following readings might be obtained if the same current was measured repeatedly using one ammeter:

2.4   2.5   2.4   2.6   2.5   2.6   2.6   2.5

Because of the uncertainty, there is variation in the last figure. To arrive at a single value for the current, you could find the mean of the above readings, and then include an estimation of the uncertainty:

current $= 2.5 \pm 0.1$

mean   uncertainty

Writing '2.5 ± 0.1' indicates that the value could lie anywhere between 2.4 and 2.6.

Note:
- On a calculator, the mean of the above readings works out at 2.5125. However, as each reading was made to only two significant figures, the mean should also be given to only two significant figures i.e. 2.5.
- Each of the above readings may also include a systematic uncertainty.

Uncertainty as a percentage

Sometimes, it is useful to give an uncertainty as a percentage. For example, in the current measurement above, the uncertainty (0.1) is 4% of the mean value (2.5), as the following calculation shows:

percentage uncertainty $= \dfrac{0.1}{2.5} \times 100 = 4$

So the current reading could be written as $2.5 \pm 4\%$.

Combining uncertainties

Sums and differences Say you have to *add* two length readings, A and B, to find a total, C. If $A = 3.0 \pm 0.1$ and $B = 2.0 \pm 0.1$, then the minimum possible value of C is 4.8 and the maximum is 5.2. So $C = 5.0 \pm 0.2$.

Now say you have to subtract B from A. This time, the minimum possible value of C is 0.8 and the maximum is 1.2 . So $C = 1.0 \pm 0.2$, and the uncertainty is the same as before.

If $C = A + B$ or $C = A - B$, then

> uncertainty = uncertainty + uncertainty
> in C in A in B

The same principle applies when several quantities are added or subtracted: $C = A + B - F - G$, for example.

Calculated results

Say you have to calculate a resistance from the following readings:

> voltage = 3.3 V   (uncertainty $\pm$ 0.1 V, or $\pm$ 3%)
> current = 2.5 A   (uncertainty $\pm$ 0.1 A, or $\pm$ 4%)

Dividing the voltage by the current on a calculator gives a resistance of 1.32 Ω. However, as the combined uncertainty is $\pm$7%, or $\pm$ 0.1 Ω, the calculated value of the resistance should be written as 1.3 Ω ($\pm$7%) or 1.3 Ω ($\pm$0.1 Ω).

Showing uncertainties on graphs

In an experiment, a wire is kept at a constant temperature. You apply different voltages across the wire and measure the current through it each time. Then you use the readings to plot a graph of current against voltage.

The general direction of the points suggests that the graph is a straight line. However, before reaching this conclusion, you must be sure that the points' scatter is due to random uncertainty in the current readings. To check this, you could estimate the uncertainty and show this on the graph using short, vertical lines called uncertainty bars. The ends of each bar represent the likely maximum and minimum value for that reading. In the example below, the **uncertainty bars** show that, despite the points' scatter, it is reasonable to draw a straight line through the origin.

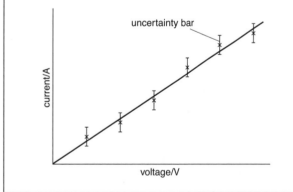

Products and quotients If $C = A \times B$ or $C = A/B$, then

> % uncertainty = % uncertainty + % uncertainty
> in C in A in B

For example, say you measure a current I, a voltage V, and calculate a resistance R using the equation $R = V/I$. If there is a 3% uncertainty in V and a 4% uncertainty in I, then there is a 7% uncertainty in your calculated value of R.

Note:
- The above equation is only an approximation – and a poor one for uncertainties greater than about 10%.
- To check that the equation works, try calculating the maximum and minimum values of C if, say, A is 100 ± 3 and B is 100 ± 4. You should find that $A \times B$ is $10\,000 \pm$ approximately 700 (i.e. 7%).
- The principle of adding % uncertainties can be applied to more complex equations: $C = A^2 B/FG$, for example. As $A^2 = A \times A$, the % uncertainty in A^2 is twice that in A.

Choosing a graph

The general equation for a straight-line graph is

$$y = mx + c$$

In this equation, m and c are **constants**, as shown below. y and 3 are **variables** because they can take different values. x is the **independent variable**. y is the **dependent variable**: its value depends on the value of x.

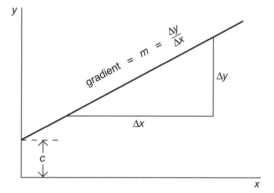

In experimental work, straight-line graphs are especially useful because the values of constants can be found from them.

Labelling graph axes Strictly speaking, the scales on the graph's axes are pure, unitless numbers and not voltages or currents. Take a typical reading:

> voltage = 10 V

This can be treated as an equation and rearranged to give:

> voltage/V = 10

That is why the graph axes are labelled 'voltage/V' and 'current/A'. The values of these are pure numbers.

Reading a micrometer

The length of a small object can be measured using a micrometer screw gauge. You take the reading on the gauge like this:

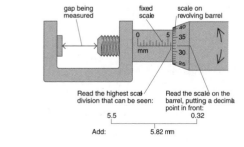

Reading a vernier

Some measuring instruments have a vernier scale on them for measuring small distances (or angles). You take the reading like this:

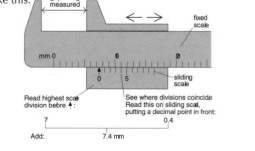

How Science Works

There are 12 aspects to 'How science works' that are included in all the AS and A level specifications. Some of the specifications have incorporated these ideas into different modules, and some have rewritten them so that the wording is different.
The 12 aspects are listed here with the original wording, together with some guidance on interpretation, and the type of questions that could be asked.

Use theories, models, and ideas to develop and modify scientific explanations

There are many historical examples of scientists making some observations and then using creative thinking and imagination to interpret the data and develop an explanation. The first step is to come up with an **idea** – an initial thought about the reasons for the observations. This is then extended and worked into a model, maybe combining several ideas.

One definition of a **model** is:

A representation of a system that allows for investigation of the properties of the system and, in some cases, prediction of future outcomes.

The model is then tested and, if it works, can be set out as a **theory**.

One definition of a scientific theory is:

A set of statements or principles that explain observations, especially a set that has been repeatedly tested or is widely accepted and can be used to make predictions about natural phenomena.

You may be asked to give an example. Here are some:

Galileo Galilei timed objects rolling down an inclined plane and concluded that falling objects accelerate. Isaac Newton suggested a model for gravity and showed that freely-falling objects have the same acceleration. He developed the theory of gravity.

Scientists often use microscopic models, such as that of particle behaviour, to explain macroscopic behaviour. For example, the conduction of electricity and the electron model.

Other examples of using models to develop theories include:

- Energy transfers for a rollercoaster
- The electron and quantization of charge
- Electrical resistance
- Young's slits
- The photoelectric effect.

At this stage in your study of science you are unlikely to be thinking up new theories, but you may be doing experiments to see if your observations fit with a model.

Use knowledge and understanding to pose scientific questions, define scientific problems, present scientific arguments and scientific ideas

As part of your course you use scientific theories to answer scientific questions or address scientific problems. In addition, you are expected to identify scientific questions or problems (within a given context). You may be presented with a hypothesis (an untested theory based on observations) or be asked to suggest one. The hypothesis needs to be tested by experiment, and if a reliable experiment does not support a hypothesis it must be changed.

When presenting arguments and ideas you should be able to

distinguish between questions that science can address, and those that science cannot address. For example, whether a view is beautiful is not a question science can answer.

A historical example is the photoelectric effect.

The question was 'For a metal that shows the photoelectric effect, why is there a threshold frequency? (Why does high intensity red light cause no emission of photoelectrons, but low intensity UV radiation does cause emission?)' Albert Einstein suggested a hypothesis based on Max Planck's ideas: that the radiation was quantized and arrived in packets called photons, with energy $E = hf$.

Use appropriate methodology, including ICT, to answer scientific questions and solve scientific problems

This includes how you conduct experimental work, for example:

- planning, or following a plan, of an investigation
- identifying the dependent and independent and control variables
- selecting appropriate apparatus and methods (including ICT) to carry out reliable experiments
- choosing instruments with appropriate sensitivity and precision (see opposite)
- justifying the methods used during experiments (including the use of ICT) to collect valid and reliable data and produce scientific theories
- using ICT (spreadsheets, for example) to develop scientific models or plot graphs, and dataloggers to monitor physical changes.

Sensitivity and precision
Take the example of measuring mass. The more *sensitive* a balance is, the smaller variation in mass the balance can detect and measure. A mass smaller than the sensitivity of a balance is not detectable using the balance.

If the mass of an object is measured many times, the *precision* is indicated by the spread of the results. If the measurements are all very close, the precision of the instrument is greater.

Accuracy and precision
A measurment with great precision is not the same as one with great accuracy, as illustrated by the diagram:

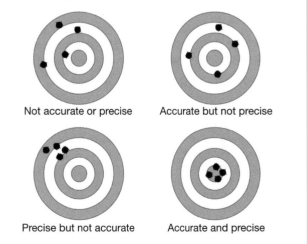

Not accurate or precise | Accurate but not precise

Precise but not accurate | Accurate and precise

Carry out experimental and investigative activities, including appropriate risk management, in a range of contexts

You should be able to show that you can:

- follow experimental procedure in a sensible order
- use appropriate apparatus and methods to make accurate and reliable measurements (see above)
- identify and minimize significant sources of experimental error
- identify and take account of risks in carrying out practical work
- produce a risk assessment before carrying out a range of practical work.

An example would be to recognize that you should use eye protection when stretching wires or springs, and ensure that heavy weights cannot cause damage to people or objects if they fall.

Analyse and interpret data to provide evidence, recognizing correlations and causal relationships

This refers to the methods you use in your experimental work, and what you do with the data you collect.

You will be expected to:

- record data in tables, and sometimes use equations to calculate values that you add to the table (for example, if you measure the extention of a wire you might add strain, ε to your table)
- plot and use graphs to establish or verify relationships between variables
- calculate the gradient and find the intercepts of straight-line graphs
- analyse data, including graphs, to identify patterns and relationships (correlation and cause, for example).

Correlation and cause
Analyse graphs of datasets that show different correlations.

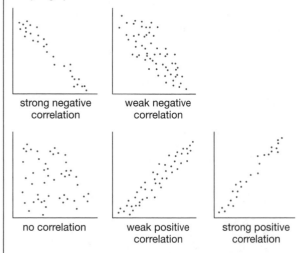

strong negative correlation   weak negative correlation

no correlation   weak positive correlation   strong positive correlation

Remember that a correlation does not necessarily show that one thing causes the other. (Children with larger feet are better at spelling but this is because older children have larger feet and are better at spelling then younger children, not because larger feet *cause* better spelling.)

Evaluate methodology, evidence and data, and resolve conflicting evidence

You should be able to:

- recognize, and distinguish between, systematic and random errors
- estimate the errors in measurements
- use data, graphs, and other experimental evidence to draw conclusions
- use the (estimated) most significant error to assess the reliability of your conclusion
- evaluate the validity of conclusions in the light of the experimental methods used
- recognize conflicting evidence.

These skills can be applied to any experiments.

Appreciate the tentative nature of scientific knowledge

In everyday speech, 'tentative' usually means hesitant or unsure, but it also means 'not fully worked out', or 'a work in progress'. This is the nature of scientific knowledge.

Once scientists find a theory that works well, they accept and use it for as long as it works, but they recognize and accept that if new observations in the future conflict with the theory, and a better explanation is offered, then the accepted theory will change. This is unlike some other types of knowledge or belief.

A good example is the way in which Newton's laws of motion, which work perfectly well for everyday speeds, do not correctly describe behaviour at speeds approaching light speed.

The modification of our theory of motion to take into account Einstein's theory of relativity was necessary.

Einstein's theory was originally theoretical, but it predicted distortion of starlight during an eclipse, which was observed in 1919. In 1971 accurate atomic clocks were available that were sensitive enough to detect the difference in the time taken by clocks flown in different directions around the world for 3 days. When you use a satellite navigation system, the processing takes into account relativity in determining your position.

Other examples of the development of theories include:

- theories of motion – Galileo and Aristotle
- the photoelectric effect.

Communicate information and ideas in appropriate ways using appropriate terminology

Using the correct scientific terminology avoids confusion. You should be able to write explanations using correct scientific terms, and support your arguments with equations, diagrams and clear sketch graphs. This applies to exam papers, practical work, and investigations.

Consider applications and implications of science and appreciate their associated benefits and risks

You may be asked to discuss the risk associated with an activity from almost any physics topic, in terms of the actual level of the risk and its potential consequences. People's perception of risk depends on factors such as how familiar the risk is. For example, people overestimate the risk of an aircraft crash, and underestimate the risk of a car crash.

There are some physics topics with more obvious benefits and risks than others.

Fossil fuels, electricity generation, and global warming

Since the Industrial Revolution, the application of physics has included many activities that involve burning fossil fuels, which release carbon dioxide. Most scientists now think that global warming is at least partly due to human activities. You should be aware of the impact this has had on the environment and how scientists are using their current findings to inform decision-makers of the consequence of global warming and advise them how to minimize its effects.

Ultraviolet radiation

Radiation reaching us from the Sun includes UVA, which has an aging effect, and UVB, which can damage the cornea of the eye and cause skin cancer. Sunlight is needed by the body to produce vitamin D, which is required for bone growth in children and has a role in reducing the risk of other cancers. Ultraviolet radiation with shorter wavelengths, UVC, is more dangerous. Sunscreen filters out the ultraviolet radiation so that it does not reach the skin.

Other topics include:
- car safety, including the global positioning system (GPS), which has benefits we use everyday in satellite navigation systems, but the system can also be used to target air strikes accurately
- material properties.

Consider ethical issues in the treatment of humans, other organisms, and the environment

You should be able to identify ethical issues arising from the application of science as it impacts on humans and the environment, and discuss scientific solutions from a range of ethical viewpoints.

Scientific research is funded by society, either through public funding or through private companies that obtain their income from commercial activities. Scientists have a duty to consider ethical issues associated with their findings. They set up groups to decide what should be permitted, and also contribute to groups set up by society to make decisions about what should be permitted.

Individual scientists have ethical codes that are often based on humanistic, moral, and religious beliefs.

Science has provided solutions to problems, but it is up to society as a whole (including scientists) to judge whether the solution is acceptable in view of the moral issues that result. Issues such as effects on the planet, and the economic and physical well-being of the living things on it, should be considered. Secure transmission of data is important if people are to be confident that personal data cannot be intercepted in transmission.

When a country is at war there may be difficult decisions for scientists to make. In the Second World War, scientists on both sides were in a race to build the first atom bomb.

Music can now be stored and reproduced to a high standard, but as a result infringement of copyright has become simple. Steps have been taken to make it more difficult to download music illegally – an example of science and technology providing solutions to the problems it creates.

Appreciate the role of the scientific community in validating new knowledge and ensuring integrity

It is important that new data, and new interpretations of data, should be critically evaluated. This is true whether they support established scientific theories or propose new theories.

Scientists communicate their findings to other scientists through journals and conferences. By sharing the findings of their research, scientists provide the scientific community with opportunities to replicate and further test their work. This can result in either confirming new explanations or refuting them.

Peer review

Some scientific journals state in the journal (and on their websites) that they are peer reviewed.

This means that the papers submitted by scientists for publication are sent for peer review before being accepted for publication. A peer is 'a person who is of equal standing with another in a group.' In this case, it is another scientist, or scientists, working in the same, or similar, field of research. Other scientists know that everything in the journal has been considered by another qualified, independent, scientist.

Note that some scientific magazines are not peer reviewed, but may use peer-reviewed articles as a source of information, as well as accepting other articles. A scientist reporting new research would usually publish in a peer-reviewed journal first.

Funding

The interests of the organizations that fund scientific research can influence the direction of that research. In some cases, the validity of the resulting claims may also be influenced.

The UK Government's leading funding agency for research and training in engineering and the physical sciences is the Engineering and Physical Science Research Council (EPSRC). Scientists submit proposals detailing the research they want to do and the equipment and staff they need.
The proposals are peer reviewed, and then all the proposals and the reviewer's comments are considered by a committee of scientists.

Almost all scientists work with the common aim of progressing scientific knowledge and understanding in a valid way and believe that accurate reporting of findings should take precedence over recognition of success of an individual. However, a disadvantage of the system could be that less ethical reviewers have the opportunity to benefit from other scientists' ideas and results, and to prevent publication and funding of their work. To prevent this, the reviewers' comments are sent to the author/researcher for comment. The system could work to exclude scientists with unusual research ideas or theories.

An example of different parts of the scientific community working together was the experimental discovery of electron diffraction which confirmed the dual nature of matter particles, first put forward by de Broglie as a hypothesis several years earlier.

Appreciate the ways in which society uses science to inform decision-making

Science influences decisions on an individual, local, national, and international level.

Scientific findings lead to new technologies, which enable advances to be made that have potential benefit for humans. However, these have to be balanced against the risks.

In practice, the scientific evidence available to decision-makers may be incomplete – scientific evidence should be considered as a whole. Decision-makers, who include government-appointed science advisers, are influenced by many things. These include their prior beliefs, their vested interests, special interest groups, public opinion, and the media, as well as by expert scientific evidence. The media and pressure groups often select parts of scientific evidence that support a particular viewpoint. This can influence public opinion, which in turn may influence decision-makers. Consequently, decision-makers may make socially and politically acceptable decisions based on incomplete evidence. The following are examples of this.

Electric cars may replace petrol vehicles if batteries are developed that give a greater range than those at present. Until then, car buyers are unlikely to be persuaded to buy electric cars.

Satellite tracking for purposes such as road pricing may be implemented without adequate trials because of pressure group influence.

The improved communication that digital electronics bring to society means that people can find out more easily what is happening and give their views. (For example, there are many petitions on the Downing Street website that you can sign online.) The range of information made available to decision-makers in industry, services, and government has increased now that information can be processed and presented using computers.

The expense of space travel is one area in which people have strong views. Some people regard it as a waste of money that could be used for building hospitals, for example. Others see the interest generated from space travel as beneficial and the technological spin-offs as worthwhile. Space travel contributes to global warming, but also to a greater understanding of climate on Earth and other planets.

Unit G481 Mechanics

Module 1: Motion

1.1.1 Physics quantities and units

Physical quantity

Say a plank is 2 metres long. This measurement is called a **physical quantity**. In this case, it is a length. It is made up of two parts:

2 m

magnitude   unit
(number)

Note:
- '2 m' really means '2 × metre', just as, in algebra, $2y$ means '2 × y'.

Numbers with no units

When something is measured it has a unit, but sometimes we use a ratio. For example efficiency is the ratio of the useful output power from an appliance, divided by the input power. Power divided by power =1 so efficiency has no units.

$$\eta = \frac{\text{output power}}{\text{input power}}$$

SI derived units

There is no SI base unit for speed. However, speed is defined by an equation (see 'Kinematics', page 23). If an object travels 12 m in 3 s,

$$\text{speed} = \frac{\text{distance travelled}}{\text{time taken}} = \frac{12 \text{ m}}{3 \text{ s}} = 4 \frac{\text{m}}{\text{s}}$$

The units m and s have been included in the working above and treated like any other numbers or algebraic quantities. To save space, the final answer can be written as 4 m/s, or 4 m s^{-1}. (Remember, in maths, $1/x = x^{-1}$ etc.)

The unit m s^{-1} is an example of a **derived SI unit**. It comes from a defining equation. There are other examples below. Some derived units are based on other derived units. And some derived units have special names. For example, 1 joule per second (J s^{-1}) is called 1 watt (W).

Angles

Angles are measured in degrees, (°). Remember that a right angle is 90° and a complete revolution is 360°. Angles can also be measured in radians.

SI base units

Scientific measurements are made using SI units (standing for Système International d'Unités). The system starts with a series of **base units**, the main ones being shown in the table above right. Other units are derived from these.

SI base units have been carefully defined so that they can be accurately reproduced using equipment available to national laboratories throughout the world.

Physical quantity	Unit	
	Name	Symbol
length	metre	m
mass	kilogram	kg
time	second	s
current	ampere	A
temperature	kelvin	K
amount*	mole	mol

* In science, 'amount' is a measurement based on the number of particles (atoms, ions or molecules) present. One mole is 6.02×10^{23} particles, a number which gives a simple link with the total mass. For example, 1 mole (6.02×10^{23} atoms) of carbon-12 has a mass of 12 grams. 6.02×10^{23} is called the **Avogadro constant**.

Physical quantity	Defining equation (simplified)	Derived unit	Special symbol (and name)
area	length^2	m^2	—
volume	length^3	m^3	—
speed	distance/time	m s^{-1}	—
acceleration	speed/time	m s^{-2}	—
force	mass × acceleration	kg m s^{-2}	N (newton)
work and energy	force × distance	N m	J (joule)
torque and moment of a force	force × perpendicular distance	N m	—
power	work/time	J s^{-1}	W (watt)
pressure	force/area	N m^{-2}	Pa (pascal)
density	mass/volume	kg m^{-3}	—
charge	current × time	A s	C (coulomb)
voltage	energy/charge	J C^{-1}	V (volt)
resistance	voltage/current	V A^{-1}	Ω (ohm)
resistivity	resistance × area/length	Ω m	—
frequency	1/period	s^{-1}	Hz (hertz)

Using base units to check equations

Each term in the two sides of an equation must always have the same units or dimensions. For example,

$$\text{work} = \text{force} \times \text{distance moved}$$
$$J = N\,m = N \quad \times \quad m$$

An equation cannot be accurate if the base units on both sides do not match. It would be like claiming that '6 apples equals 6 oranges'.

Base units are a useful way of checking that an equation is reasonable.

Example *Check whether the equation $E_p = mgh$ is dimensionally correct.*

To do this, start by working out the base units of the right-hand side:

mgh units $= \text{kg} \times \text{m s}^{-2} \times \text{m} = \text{kg m}^2\text{s}^{-2}$

These are the base unit of work, and of energy.
So the equation is dimensionally correct.

Note:
- A units check cannot tell you whether an equation is accurate. For example, both of the following are dimensionally correct, but only one is right:

$$E_p = mgh \qquad E_p = 2mgh$$

Prefixes

Prefixes can be added to SI base and derived units to make larger or smaller units.

Prefix	Symbol	Value	Prefix	Symbol	Value
pico	p	10^{-12}	kilo	k	10^3
nano	n	10^{-9}	mega	M	10^6
micro	μ	10^{-6}	giga	G	10^9
milli	m	10^{-3}	tera	T	10^{12}
centi	c	10^{-2}			

For example,

$1\ \text{mm} = 10^{-3}\ \text{m} \qquad 1\ \text{km} = 10^3\ \text{m}$

Note:
- 1 gram (10^{-3} kg) is written '1 g' and not '1 mkg'.

Estimates of size

To estimate a reasonable size for one of the physical quantities listed, remember a few typical values and work out from those whether the quantity is bigger or smaller, and how many times bigger or smaller.

Example: *Force*

A small apple has a weight of about 1 N. It has a mass of about 100 g.

An 8 stone person has a weight of about 500 N.

An estimate for the tension in the lamp flex (the weight of the lamp) is 50 N.

Example: *Speed*

A car travelling at 70 mph has a speed of about 30 m/s.

The speed of sound in air is about 330 m s^{-1}.

The speed of a jet aircraft is closer to that of sound than that of a car, but usually less than sound, so an estimate is 250 m s^{-1}.

Here are some useful values to help with estimating:

	Mass/kg		Time/s
Sun	2×10^{30}	Age of the Universe	4.3×10^{17}
Earth	6×10^{24}	One year	3.2×10^7
Car	1200		
Proton	1.6×10^{-27}		
Electron	9×10^{-31}		

	Length/m		Temperature/K
Diameter of Milky Way galaxy	10×10^{20}	Room temperature	300
Earth radius	6.4×10^6	Surface of the Sun	5800
Height of a house	10		
Diameter of a carbon atom	1.5×10^{-10}		
Diameter of a proton	1.6×10^{-15}		

1.1.2 Scalars and vectors

Scalars and vectors

Vectors are quantities that have magnitude and direction. Examples of these in this section are displacement, velocity, and acceleration. Force is also a vector. When adding vectors you must allow for their direction. Two 6 N forces are being added below. In one case the resultant is 12 N. In the other it is zero.

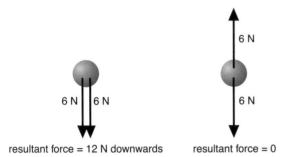

resultant force = 12 N downwards   resultant force = 0

Scalars are quantities with magnitude only. Examples are distance and speed and mass. Scalar addition is simple: if 6 kg of mass is added to 6 kg of mass the total is always 12 kg. Another example of a scalar quantity is energy. If an object has 6 J of potential energy and 6 J of kinetic energy, the total energy is 12 J.

Vector arrows

Vectors are quantities which have both magnitude (size) and direction. Examples include displacement and force.

For problems in one dimension (e.g. vertical motion), vector direction can be indicated using + or −. But where two or three dimensions are involved, it is often more convenient to represent vectors by arrows, with the length and direction of the arrow representing the magnitude and direction of the vector. The arrowhead can either be drawn at the end of the line or somewhere else along it, as convenient. Here are two displacement vectors.

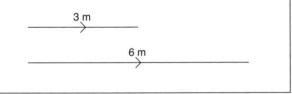

Calculating the resultant

When the two vectors are at right angles, for example the 3 N and 4 N forces on the right, the resultant can be calculated as an alternative to drawing a scale diagram.

The magnitude of the resultant force is given by Pythagoras' theorem:

$F = \sqrt{[(3\ \text{N})^2 + (4\ \text{N})^2]} = 5\ \text{N}$

The direction is given by $\tan \theta = \dfrac{3\ \text{N}}{4\ \text{N}} = 0.75$

Where θ is the angle the resultant makes with the 4 N force. $\theta = 37°$

Adding vectors

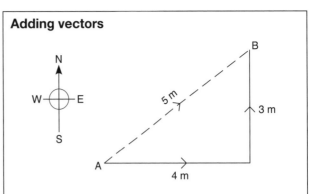

If someone starts at A, walks 4 m East and then 3 m North, they end up at B, as shown above. In this case, they are 5 m from where they started. It can be shown by drawing a scale diagram. This is an example of vector addition. Two displacement vectors, of 3 m and 4 m, have been added to produce a *resultant* – a displacement vector of 5 m.

This principle works for any type of vector. Below, forces of 3 N and 4 N act at right-angles through the same point, O. The *triangle of vectors* gives their resultant. Starting from point O the 4 N force is drawn to scale, followed by the 3 N force, or the 3 N force followed by the 4 N force, the order does not matter as long as the arrows form a continuous path

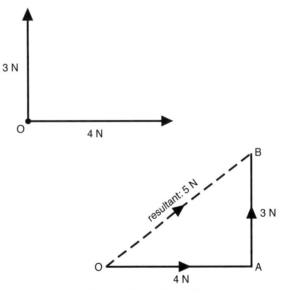

from O to A to B. The resultant is then OB.

The vectors do not have to be at right angles. In the following example the wind is blowing at an angle of 40° to the direction the aircraft is flying, so that its resultant velocity is found by drawing the vector triangle XYZ.

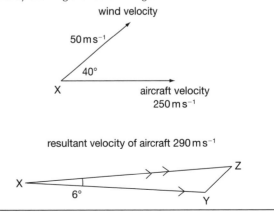

Components

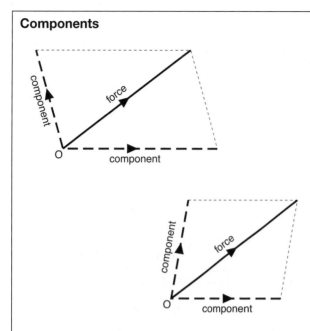

In working out the effects of a force (or other vector), the most useful components to consider are those at right-angles, as in the following example.

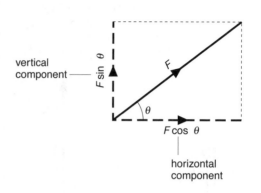

Two forces acting through a point can be replaced by a single force (the resultant) which has the same effect. Conversely, a single force can be replaced by two forces which have the same effect – a single force can be **resolved** into two **components**. Two examples of the components of a force are shown above, though any number of other sets of components is possible.

Note:
- Any vector can be resolved into components.
- The components above are shown as dashed lines to remind you that they are a *replacement* for a single force. There are *not* three forces acting.

Below, you can see why the horizontal and vertical components have magnitudes of $F \cos \theta$ and $F \sin \theta$.

$$\cos \theta = \frac{F_x}{F}$$

So $\quad F_x = F \cos \theta$

$$\sin \theta = \frac{F_y}{F}$$

So $\quad F_y = F \sin \theta$

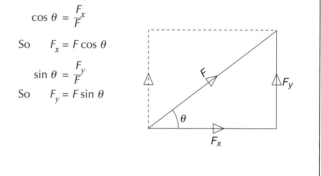

Resolving problem

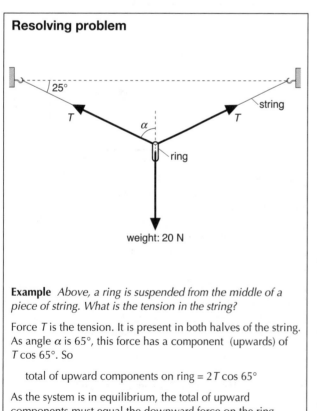

Example *Above, a ring is suspended from the middle of a piece of string. What is the tension in the string?*

Force T is the tension. It is present in both halves of the string. As angle α is 65°, this force has a component (upwards) of $T \cos 65°$. So

    total of upward components on ring $= 2T \cos 65°$

As the system is in equilibrium, the total of upward components must equal the downward force on the ring.

So $\quad 2T \cos 65° = 20$

This gives $\quad T = 24$ N

1.1.3 Kinematics

Displacement

Displacement is distance moved in a particular direction. It is a vector quantity.

The SI unit of displacement is the *metre* (m).

A ———— 12 m ————→ B

The arrow above represents the displacement of a particle which moves 12 m from A to B. However, with horizontal or vertical motion, it is often more convenient to use a '+' or '−' to show the vector direction. For example,

Movement of 12 m *to the right*: displacement = +12 m
Movement of 12 m *to the left*: displacement = −12 m

Displacement is not necessarily the same as distance travelled. For example, when the ball below has returned to its starting point, its vertical displacement is zero. However, the distance travelled is 10 m.

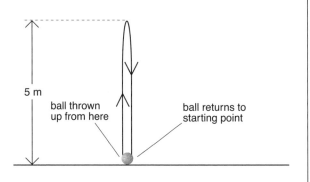

Average speed and instantaneous speed

Speed is distance travelled per unit time.

It is a scalar quantity and the SI unit is metre/second which is abbreviated to $m\,s^{-1}$.

If the speed of an object is changing during the time that the speed is measured the calculated speed will be an average value, so we call this the average speed.

Average speed is the distance travelled divided by the time taken.

Instantaneous speed is the speed at a particular instant in time.

As the time interval over which the speed is measured gets smaller and smaller the value for the speed approaches the instantaneous speed. On a graph of distance travelled against time taken, instantaneous speed is the gradient of the line.

Speed and velocity

Average speed is calculated like this:

$$\text{average speed} = \frac{\text{distance travelled}}{\text{time taken}}$$

For example, if an object travels 12 m in 2 s, its average speed is $6\,m\,s^{-1}$.

Velocity is displacement per unit time.

Average velocity is calculated like this:

$$\text{average velocity} = \frac{\text{displacement}}{\text{time taken}}$$

Like speed it is measured in $m\,s^{-1}$, but it is a vector quantity and also has a direction. For example, $3\,m\,s^{-1}$ NE.

The velocity vector above is for a particle moving to the right at $6\,m\,s^{-1}$. However, as with displacement, it is often more convenient to use a '+' or '−' for the vector direction.

Average velocity is not necessarily the same as average speed. For example, if a ball is thrown upwards and travels a total distance of 10 m before returning to its starting point 2 s later, its average speed is $5\,m\,s^{-1}$. But its average velocity is zero, because its displacement is zero.

Upwards and downwards

A ball bounces upwards from the ground. The graph on the right shows how the velocity of the ball changes from when it leaves the ground until it hits the ground again. Downward velocity has been taken as positive. Air resistance is assumed to be negligible.

Initially, the ball is travelling upwards, so it has negative downward velocity. This passes through zero at the ball's highest point and then becomes positive.

The gradient of the graph is constant and equal to *g*.

Note:
- The ball has downward acceleration *g*, even when it is travelling upwards. (Algebraically, losing upward velocity is the same as gaining downward velocity.)
- The ball has downward acceleration *g*, even when its velocity is zero (at its highest point).

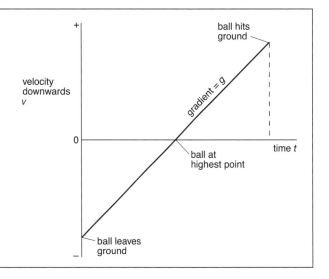

Acceleration

Acceleration = rate of change of velocity.

Average acceleration is calculated like this:

$$\text{average acceleration} = \frac{\text{change in velocity}}{\text{time taken}}$$

The SI unit of acceleration is the m s^{-2} (sometimes written m/s^2). For example, if an object gains 6 m s^{-1} of velocity in 2 s, its average acceleration is 3 m s^{-2}.

Acceleration is a vector. The acceleration vector above is for a particle with an acceleration of 3 m s^{-2} to the right. However, as with velocity, it is often more convenient to use a '+' or '−' for the vector direction.

If velocity *increases* by 3 m s^{-1} every second, the acceleration is +3 m s^{-2}. If it *decreases* by 3 m s^{-1} every second, the acceleration is −3 m s^{-2}.

Mathematically, an acceleration of −3 m s^{-2} *to the right* is the same as an acceleration of +3 m s^{-2} *to the left*.

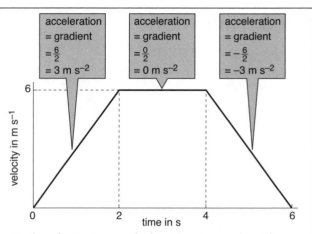

On the velocity–time graph above, you can work out the acceleration over each section by finding the *gradient* of the line. The gradient is calculated like this:

$$\text{gradient} = \frac{\text{gain along } y\text{-axis}}{\text{gain along } x\text{-axis}}$$

Displacement–time graphs

Uniform velocity The graph below describes the motion of a car moving with uniform velocity. The displacement and time have been taken as zero when the car passes a marker post. The gradient of the graph is equal to the velocity v:

$$v = \frac{\varnothing s}{\varnothing t}$$

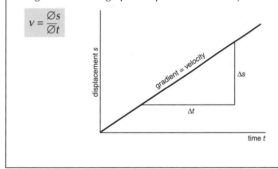

Changing velocity The gradient of this graph is increasing with time, so the velocity is increasing. The velocity v at any instant is equal to the gradient of the *tangent* at that instant.

In calculus notation $\quad v = \dfrac{ds}{dt}$

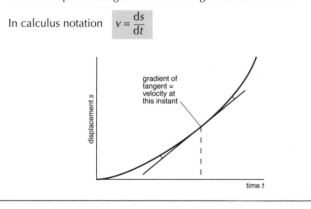

Velocity–time graphs

The graphs which follow are for three examples of *linear* motion (motion in a straight line).

Graph A (right) shows how the velocity of a stone would change with time, if the stone were dropped near the Earth's surface and there were no air resistance to slow it.
The stone has a *uniform* (unchanging) acceleration *a* which is equal to the gradient of the graph:

$$a = \frac{\varnothing v}{\varnothing t}$$

In this case, the acceleration is *g* (9.81 m s^{-2}).

If air resistance is significant, then the graph is no longer a straight line (see 'Non-linear motion' page 31).

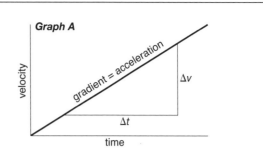

Graph B below is for a car travelling at a steady velocity of 30 m s^{-1}. In 2 s, the car travels a distance of 60 m. Numerically, this is equal to the area under the graph between the 0 and 2 s points. (Note: the area must be worked out using the scale numbers, not actual lengths.)

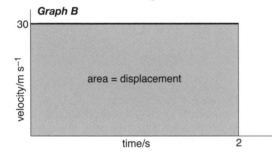

Graph C below is for a car with a changing velocity. However, the same principle applies as before: the area under the graph gives the distance travelled. (This is also true if the graph is not a straight line.)

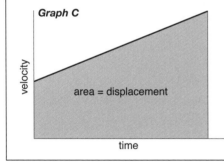

Uniform acceleration The graph below describes the motion of a car gaining velocity at a steady rate. The time has been taken as zero when the car is stationary. The gradient of the graph is equal to the acceleration.

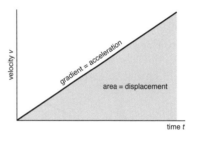

Changing acceleration The acceleration *a* at any instant is equal to the gradient of the *tangent* at that instant.

In calculus notation $\quad a = \dfrac{\mathrm{d}v}{\mathrm{d}t}$

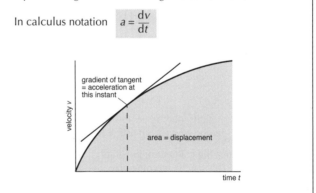

1.1.4 Linear motion

Equations of motion

The car below has uniform acceleration. In the following analysis, only motion between X and Y will be considered.

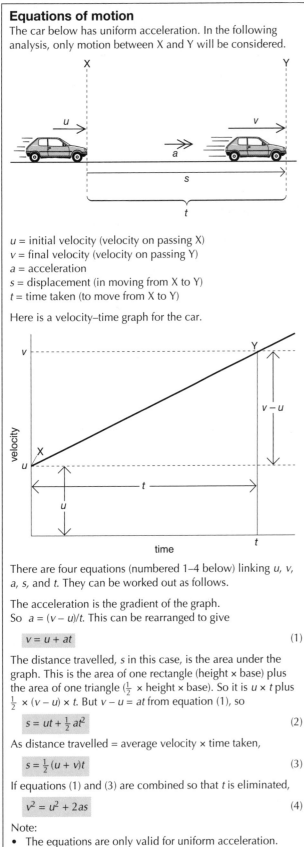

u = initial velocity (velocity on passing X)
v = final velocity (velocity on passing Y)
a = acceleration
s = displacement (in moving from X to Y)
t = time taken (to move from X to Y)

Here is a velocity–time graph for the car.

There are four equations (numbered 1–4 below) linking u, v, a, s, and t. They can be worked out as follows.

The acceleration is the gradient of the graph.
So $a = (v - u)/t$. This can be rearranged to give

$$v = u + at \tag{1}$$

The distance travelled, s in this case, is the area under the graph. This is the area of one rectangle (height × base) plus the area of one triangle ($\frac{1}{2}$ × height × base). So it is $u \times t$ plus $\frac{1}{2} \times (v - u) \times t$. But $v - u = at$ from equation (1), so

$$s = ut + \tfrac{1}{2}at^2 \tag{2}$$

As distance travelled = average velocity × time taken,

$$s = \tfrac{1}{2}(u + v)t \tag{3}$$

If equations (1) and (3) are combined so that t is eliminated,

$$v^2 = u^2 + 2as \tag{4}$$

Note:
- The equations are only valid for uniform acceleration.
- Each equation links a different combination of factors. You must decide which equation best suits the problem you are trying to solve.
- You must allow for vector directions. With horizontal motion, you might decide to call a vector to the right positive (+). With vertical motion, you might call a downward vector positive. So, for a stone thrown upwards at 30 m s^{-1}, $u = -30$ m s^{-1} and $g = +9.81$ m s^{-2}.

The motion of falling bodies

The gravitational force on objects causes them to accelerate as they fall.

Near the Earth's surface the acceleration of free fall, g is 9.81 m s^{-2}

When there is negligible air resistance or drag the equations of motion for constant acceleration in a straight line can be used:

(For simplicity, units are not shown in some equations.)

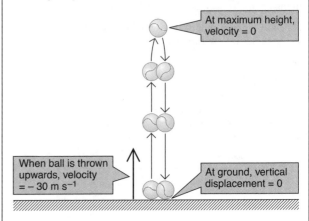

At maximum height, velocity = 0

When ball is thrown upwards, velocity $= -30$ m s^{-1}

At ground, vertical displacement = 0

Example 1 *A ball is thrown upwards at 30 m s^{-1}. What time will it take to reach its highest point?*
The ball's motion only needs to be considered from when it is thrown to when it reaches its highest point. These are the 'initial' and 'final' states in any equation used.

When the ball is at it highest point, its velocity v will be zero. So, taking downward vectors as positive,

$$u = -30 \text{ m s}^{-1} \quad v = 0 \quad a = g = 9.81 \text{ m s}^{-2} \quad t \text{ is to be found.}$$

In this case, an equation linking u, v, a, and t is required. This is equation (1) on the opposite page:

$$v = u + at$$

So $0 = -30 + 9.81t$

Rearranged, this gives $t = 3.1$ s.

Example 2 *A ball is thrown upwards at 30 m s^{-1}. What is the maximum height reached?*

In this case,

$$u = -30 \text{ m s}^{-1} \quad v = 0 \quad a = g = 9.81 \text{ m s}^{-2} \quad s \text{ is to be found.}$$

This time, the equation required is (4) on the opposite page:

$$v^2 = u^2 + 2as$$

So $0 = (-30)^2 + (2 \times 9.81 \times s)$

This gives $s = -46$ m.

(Downwards is positive, so the negative value of s indicates an *upward* displacement.)

Example 3 *A ball is thrown upwards at 30 m s^{-1}. For what time is it in motion before it hits the ground?*

When the ball reaches the ground, it is back where it started, so its displacement s is zero. Therefore

$$u = -30 \text{ m s}^{-1} \quad s = 0 \quad a = g = 9.81 \text{ m s}^{-2} \quad t \text{ is to be found.}$$

This time, the equation required is (2) on the opposite page:

$$s = ut + \tfrac{1}{2}at^2$$

So $0 = (-30t) + (\tfrac{1}{2} \times 9.81 \times t^2)$

This gives $t = 6.1$ s.

(There is also a solution $t = 0$, indicating that the ball's displacement is also zero at the instant it is thrown.)

Theories of motion

Aristotle's ideas

Almost 2400 years ago, the Greek philosopher Aristotle developed theories of Physics that were very different to the ones we use today. For example, he said that arrows move forward after they have left the bowstring because as they move through the air they create an empty space behind them that propels them forward. He said that things could move in a substance only where they could create an empty space. The 'thinner' the substance the faster things would move, and a vacuum was impossible, because things would move at infinite speed.

To explain gravity he said that some things naturally belong to the centre of the Earth and would try to return there. However, other things, such as steam and gases, naturally belong to, and try to return to, the heavenly spheres.

These ideas lasted for almost 2000 years, although a number of Persian and European scientists showed that they did not work.

Galileo's ideas

About 400 years ago, the Italian scientist Galileo Galilei made a large number of scientific observations. He pioneered the use of standard units of length and time, so that his results could be repeated and checked. This was one of the reasons he was able to convince people that Aristotle's theories were wrong.

For example, Galileo said that two cannonballs, a light one and a heavy one, dropped from the tower of Pisa would reach the ground together, and this would be true of any objects as long as air resistance was negligible. This would have been too difficult for him to measure accurately. However, he built and used inclined planes to roll balls down, timing them to show that their acceleration did not depend on their mass, providing friction was negligible. Today in Florence you can still see the equipment he used for these experiments.

When an astronaut on the Moon dropped a feather and a hammer at the same time, because there was no air resistance they both reached the surface together. The same result can be shown on Earth if they are dropped in a vacuum tube.

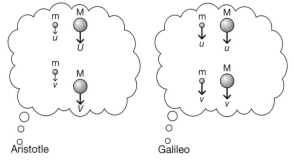

Newton's ideas

Isaac Newton continued to extend and improve our theories of motion and gravity. Today we know that a force is required to change the motion of an object, and that resistive forces such as air resistance and friction slow down moving objects. Newton's laws have been tested and shown to work on the Moon and in space.

Measuring g

By measuring the time t it takes an object to fall through a measured height h, a value of g can be found (assuming that air resistance is negligible).

In the diagram on the right, $u = 0$ $a = g$ $s = h$
Applying equation (2) on the opposite page gives

$$s = ut + \tfrac{1}{2}at^2$$

So $h = 0t + \tfrac{1}{2}gt^2$

This gives $g = \dfrac{2h}{t^2}$.

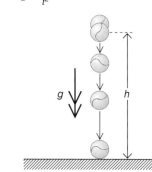

To measure the time accurately it is best to use a method of automatically starting a timer when the ball is released, and switching it off when it reaches the ground. The diagram shows a set-up where a steel ball is held by an electromagnet. When the switch is pressed the electromagnet is switched off and the timer starts. When the ball hits the trapdoor it breaks the circuit and stops the timer.

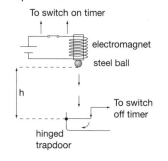

- Set up as shown.
- Arrange the timer to start when the electromagnet is switched off and stop when the hinged trapdoor opens.
- Measure h from the bottom of the ball to the trapdoor.
- Measure the time to fall 3 times and take the average.
- Repeat the experiment for a range of heights between 0.5 m and 2 m.
- Plot a graph of 2 h against t^2.
- Your graph should be a straight line with gradient g.

Projectiles

A projectile is an object that:

- has sideways (horizontal) movement as well as falling under gravity, but is not powered and so has no horizontal acceleration

- has negligible air resistance, which can be ignored.

Examples are balls and bullets.

Downwards and sideways

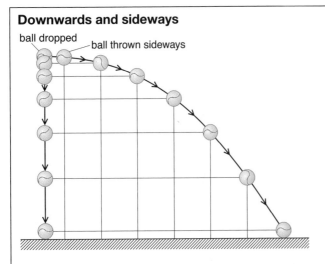

ball dropped — ball thrown sideways

Above, one ball is dropped, while another is thrown sideways at the same time. There is no air resistance. The positions of the balls are shown at regular time intervals.

- Both balls hit the ground together. They have the same downward acceleration g.
- As it falls, the second ball moves sideways over the ground at a steady speed.

Results like this show that the vertical and horizontal motions are independent of each other.

Example *Below, a ball is thrown horizontally at 40 m s^{-1}. What horizontal distance does it travel before hitting the water? (Assume air resistance is negligible and $g = 9.81$ m s^{-2}.)*

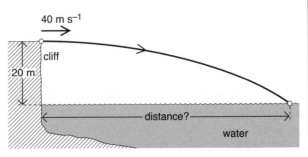

First, work out the time the ball would take to fall vertically to the sea. This can be done using the equation $s = ut + \frac{1}{2}at^2$, in which $u = 0$, $s = 20$ m, $a = -9.81$ m s^{-2}, and t is to be found. This gives $t = 2.0$ s.

Next, work out how far the ball will travel horizontally in this time (2 s) at a steady horizontal speed of 40 m s^{-1}.

As distance travelled = average speed × time, horizontal distance travelled = $40 \times 2.0 = 80$ m.

Module 2: Forces in action
1.2.1 Force

Force
Force is a vector. The SI unit is the **newton** (**N**).

If two or more forces act on something, their combined effect is called the **resultant force**.

A resultant force acting on a mass causes an acceleration. The acceleration is in the same direction as the force. The force, mass, and acceleration are linked like this:

resultant force = mass × acceleration $F = ma$

A 1 N resultant force gives a 1 kg mass an acceleration of 1 m s^{-2}.

The more mass something has, the more force is needed to produce any given acceleration.

When balanced forces act on something, its acceleration is zero. This means that it is *either* stationary *or* moving at a steady velocity (steady speed in a straight line).

The Special Theory of Relativity
Einstein rejected the idea of absolute motion through space: motion could only be relative to the observer's **frame of reference**. He developed his **special theory of relativity** (1905) from the following two postulates (assumptions):

1 Physical laws (e.g. the laws of motion) are the same in all *inertial* (unaccelerated) frames of reference.

2 The speed of light in a vacuum has the same measured value in all inertial frames of reference.

From these postulates, Einstein deduced that length and time measurements could not be absolute. They must depend on the relative motion of the observer. When time frame B has a velocity, v, relative to time from A, the mass of an object in B as observed from A depends on v.

when $v = 0$ the mass is called the rest mass, m_0

As v increases, m increases so as an object approaches the speed of light the equation $F = ma$ cannot be used because m increases.

When $v = c$, m is infinite, and so is the E_k. This effectively makes the speed of light a universal speed limit.

Calculations of force and motion
Example

A car of mass 1200 kg travels at 20 m s^{-1} brakes to a stop in 4 s. Find a) the average deceleration b) the average braking force c) the distance travelled while braking to a stop.

a) a is to be found. $u = 20$ m s^{-1}, $t = 4$ s, $v = 0$

using $v = u + at$

$0 = (20$ m s$^{-1}) + a \times (4$ s$)$

$a = -\dfrac{20 \text{ m s}^{-1}}{4 \text{ s}} = -5$ m s^{-2}

(the acceleration is negative because it is in the opposite direction to the velocity, the car is slowing down)

b) F is to be found. $m = 1200$ kg, $a = -5$ m s^{-2}

using $F = ma$

$F = (1200$ kg$) \times (-5$ m s$^{-2}) = -6000$ N

(the force is negative because it is in the opposite direction to the direction of motion)

c) s is to be found. $u = 20$ m s^{-1}, $t = 4$ s, $v = 0$, $a = -5$ m s^{-2}

using $s = ut + \frac{1}{2}at^2$

$s = (20$ m s$^{-1}) \times (4$ s$) + \frac{1}{2}(-5$ m s$^{-2}) \times (4$ s$)^2$

$s = 80$ m $- \frac{1}{2} \times 5 \times 16$ m $= 40$ m

1.2.2 Non-linear motion

Drag

When objects move through a fluid – a gas or a liquid – there is a resistive force that acts to slow the object down. It is a frictional force between the object and the fluid. This is called drag (or sometimes air resistance, if the motion is in air). The force depends on what fluid the object is moving through, the shape of the object and the velocity of the object through the fluid (or the velocity squared in many cases).

Factors that affect drag

Drag increases:

- As the speed of the object through the fluid increases. This has a big effect, drag usually increases with speed squared. So double the speed gives four times the drag.
- As the cross-sectional area of the object increases.
- As the roughness of the object increases
- For less streamlined objects. When an object is streamlined the fluid flows easily round it with no turbulence, so there is less drag.

Increasing drag

Parachutes have a large cross-sectional area increase the drag force to slow objects down. They are used on some jet aircraft when landing, and on some racing cars.

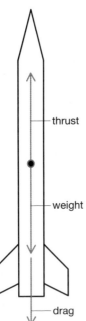

Reducing drag

When a fluid flows in layers that slide over each other so that the movement of the layers can be represented by smooth lines that do not cross the flow is said to be streamlined. Drag on cars and other objects can be reduced by using a streamlined design.

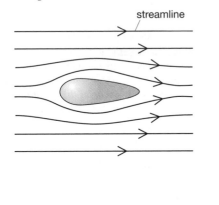

streamline

Acceleration of an object with drag

Example

This diagram shows the forces on a model rocket, fired vertically upwards using compressed air.

The mass of the rocket is 0.20 kg. The initial thrust is 4.0 N. After 2.0 s the thrust is 3.0 N and the drag force 0.80 N. What is a) the initial acceleration? b) the acceleration after 20 s?

Using $W = mg$

Weight $W = 0.20$ kg $\times$ 9.81 m s^{-2}

$W = 1.96$ N

The motion is upwards, so take up as the positive direction.

The resultant upward vertical force on the rocket = $T - W - D$

a) At the moment it is launched the speed is 0 m s^{-1}, so the drag = 0 N

Initial resultant force = 4.0 N – 1.96 N – 0 N = 2.04 N

$F = ma$

2.04 N = 0.20 kg $\times$ a

initial acceleration = 10.2 m s^{-2} = 10 ms^{-2} (2sf)

b) After 2 s resultant force = 3.0 N – 1.96 N – 0.80 N = 0.24 N

$F = ma$

0.24 N = 0.20 kg $\times$ a

After 2.0 s acceleration = 1.2 m s^{-2}

thrust

weight

drag

Example

A skydiver with her parachute has a total mass of 60 kg. She is falling with a velocity of 50 m s^{-1} when she opens her parachute. The drag increases to 700 N, what is her acceleration?

$W = ma$

$W = 60$ kg $\times$ 9.81 m s^{-2}

$W = 589$ N

Down is the direction of motion, so take down as the positive direction

Resultant downward force = $W - D = 589$ N – 700 N = –111 N

Remember that frictional forces like drag and air resistance always act in the opposite direction to motion. The negative sign shows that the resultant force is upwards.

$F = ma$

–111 N = 60 kg $\times$ a

$a = -1.85$ m s^{-2} = 1.9 m s^{-2} (2sf)

the acceleration is 1.9 m s^{-2} in an upward direction, showing that the skydiver is slowing down.

Weight and *g*

On Earth, everything feels the downward force of gravity. As for other forces, its SI unit is the newton (N).

> The weight is the gravitational force acting on an object.

Near the Earth's surface, the gravitational force on each kg is about 9.81 N: the **gravitational field strength** is 9.81 N kg^{-1}.

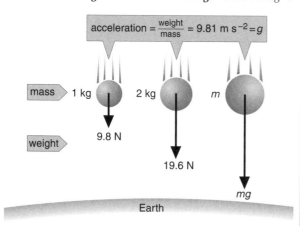

$$\text{acceleration} = \frac{\text{weight}}{\text{mass}} = 9.81 \text{ m s}^{-2} = g$$

mass: 1 kg   2 kg   *m*

weight: 9.8 N   19.6 N   *mg*

Earth

This is represented by the symbol *g*.
In the diagram above, all the masses are falling freely (gravity is the only force acting). From $F = ma$, it follows that all the masses have the same downward acceleration, *g*. This is the **acceleration of free fall**.

The force F = weight W so $W = mg$

You can think of *g*

either as a gravitational field strength of 9.81 N kg^{-1}

or as an acceleraton of free fall of 9.81 m s^{-2}.

The value $g = 9.81$ m s^{-2} is given in the data, formulae and relationships booklet, so you should always use this value. It is OK to use 9.8 m s^{-2}, but you will be penalised for using 10 m s^{-2}.

The fall of a skydiver

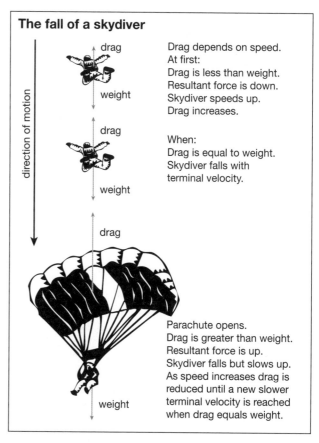

direction of motion

drag
weight

Drag depends on speed.
At first:
Drag is less than weight.
Resultant force is down.
Skydiver speeds up.
Drag increases.

drag
weight

When:
Drag is equal to weight.
Skydiver falls with terminal velocity.

drag
weight

Parachute opens.
Drag is greater than weight.
Resultant force is up.
Skydiver falls but slows up.
As speed increases drag is reduced until a new slower terminal velocity is reached when drag equals weight.

Terminal velocity

Air resistance on a falling object can be significant. As the velocity increases, the air resistance increases, until it eventually balances the weight. The resultant force is then zero, so there is no further gain in velocity. The object has reached its **terminal velocity**.

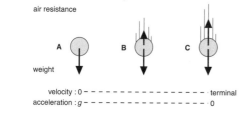

air resistance

A   B   C

weight

velocity : 0 – – – – – – – – – – – – – – – terminal
acceleration : *g* – – – – – – – – – – – – – – – 0

An object falling at terminal velocity loses potential energy but does not gain kinetic energy. Instead, the energy heats up the object and the surrounding fluid. Spacecraft and asteroids entering the Earth's atmosphere at very high speeds can burn up, so the space shuttle was covered with special heat-resistant tiles.

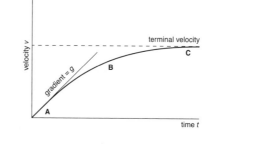

velocity *v*

terminal velocity

gradient = *g*

A   B   C

time *t*

1.2.3 Equilibrium

Equilibrium

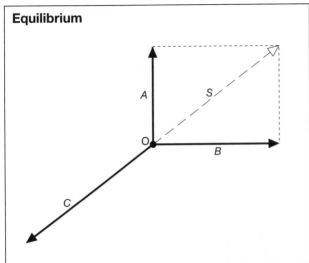

The particle O above has three forces acting on it – A, B, and C. Forces A and B can be replaced by a single force S. As force C is equal and opposite to S, the resultant of A, B, and C, is zero. This means that the three forces are in balance – the system is in **equilibrium**.

If three forces are in equilibrium, they can be represented by the three sides of a triangle, as shown below. Note that the sides and angles match those in the previous force diagram. The forces can be drawn in any order, provided that the head of each arrow joins with the tail of another.

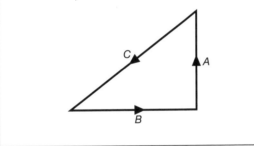

Finding the centre of gravity

To find the centre of gravity of a laminar (flat) object, suspend the object so it can swing freely, as shown in the diagram. It will come to rest so that the centre of gravity is directly below the pivot point X.

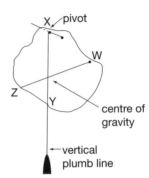

Use a plumb line to mark the vertical line from the pivot on the object, XY. The centre of gravity is on this line.

Suspend the object from a second point. Draw the second vertical line WZ. The centre of gravity will be where the two lines cross.

Conditions for equilibrium

There are two types of motion: **translational** (from one place to another) and **rotational** (turning). If a static, rigid object is in equilibrium, then

- the forces on it must balance, otherwise they would cause translational motion,
- the moments must balance, otherwise they would cause rotational motion.

The balanced beam on the opposite page is a simple system in which the forces are all in the same plane. A **coplanar** system like this is in equilibrium if

- the vertical components of all the forces balance,
- the horizontal components of all the forces balance,
- the moments about any axis balance.

To check for equilibrium, components can be taken in any two directions. However, vertical and horizontal components are often the simplest to consider. The balanced beam is especially simple because there are no horizontal forces.

Centre of gravity

All the particles in an object have weight. The weight of the whole object is the resultant of all these tiny, downward gravitational forces.

> The **centre of gravity** of an object is the point where the entire weight of an object appears to act.

For a rectangular beam with an even weight distribution, the centre of gravity is in the middle. Unless negligible, the weight must be included when analysing the forces and moments acting on the beam.

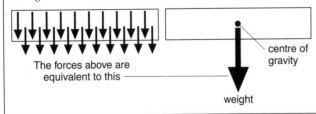

The forces above are equivalent to this

centre of gravity

weight

For a dumb-bell with uneven weight distribution, the centre of gravity is at point O because the two moments about O are equal and opposite.

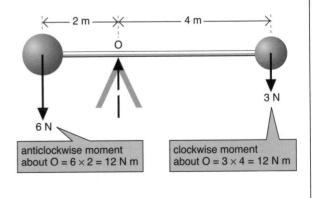

anticlockwise moment about O = 6 × 2 = 12 N m

clockwise moment about O = 3 × 4 = 12 N m

Moment of a force

The turning effect of a force about a point is called its *moment*.

moment of force = force × perpendicular distance from
about a point   point measured from the line
              of action of the force

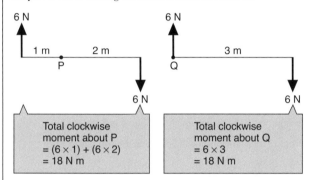

Moment of force F about O
    $= Fd$

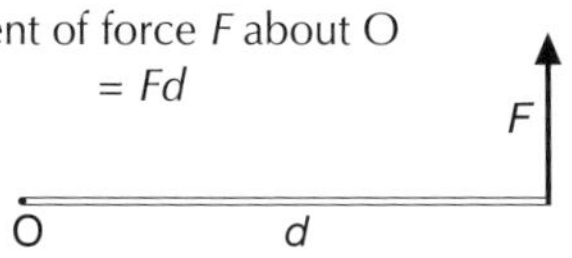

Moment of force F about O
    $= Fx$ $= Fd \cos \theta$

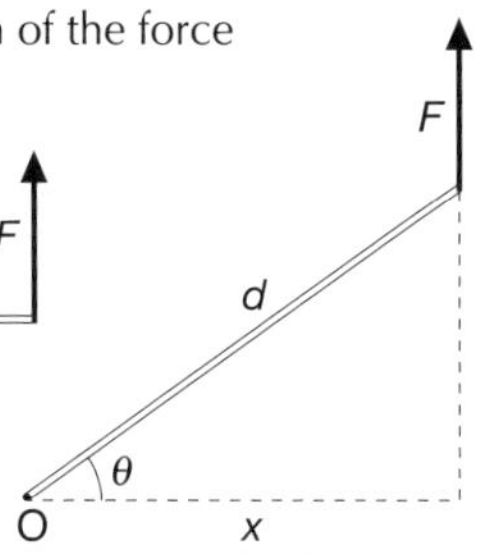

Note:

• In the diagram on the left, although O is shown as a point, it is really an *axis* going perpendicularly into the paper.

• Moments are measured in N m. However this is not the same unit as the N m, or J (joule), used for measuring energy.

• A moment can be *clockwise* or *anticlockwise*, depending on its *sense* (direction of turning). This can be indicated with a + or –. For example,

anticlockwise moment of 2 N m = +2 N m
clockwise moment of 2 N m   = –2 N m

Couples and torque

A pair of equal but opposite forces, as below, is called a *couple*. It has a turning effect but no resultant force.

6 N   6 N

1 m   2 m   3 m
   P   Q

         6 N   6 N

Total clockwise
moment about P
$= (6 \times 1) + (6 \times 2)$
$= 18$ N m

Total clockwise
moment about Q
$= 6 \times 3$
$= 18$ N m

To find the total moment of a couple, you could choose any axis, work out the two moments and add them up. Whichever axis you choose, the answer is the same, so the simplest way of calculating the total moment is like this:

moment of couple = one force × perpendicular distance between forces

Note:

• The total moment of a couple is called a *torque*.
• Strictly speaking, a couple is any system of forces which has a turning effect only i.e. one which produces rotational motion without translational (linear) motion.

Principle of moments

The beam in the diagram on the right has weights on it. (The beam itself is of negligible weight.) The total weight is supported by an upward force R from the fulcrum.

The beam is in a state of balance. It is in equilibrium.

As the beam is not tipping to the left or right, the turning effects on it must balance. So, when moments are taken about O, as shown, the total clockwise moment must equal the total anticlockwise moment. (Note: R has zero moment about O because its distance from O is zero.)

As the beam is static, the upward force on it must equal the total downward force. So $R = 10 + 8 + 4 = 22$ N.

The beam is not turning about O. But it is not turning about any other axis either. So you would expect the moments about *any* axis to balance. This is exactly the case, as you can see in the next diagram. The beam and weights are the same as before, but this time, moments have been taken about point P instead of O. (Note: R does have a moment about P, so the value of R must be known before the calculation can be done.)

The examples shown on the right illustrate the *principle of moments*, which can be stated as follows:

If an object is in equilibrium, the sum of the clockwise moment about any axis is equal to the sum of the anticlockwise moments.

Here is another way of stating the principle. In it, moments are regarded as + or –, and the *resultant moment* is the algebraic sum of all the moments:

If a rigid object is in equilibrium, the resultant moment about any axis is zero.

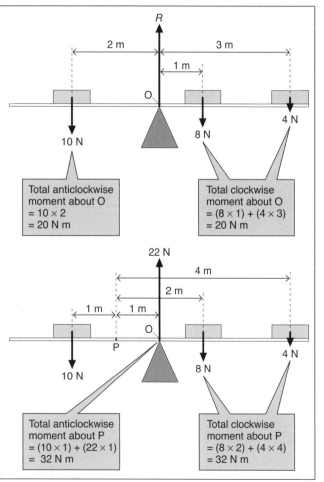

Total anticlockwise
moment about O
$= 10 \times 2$
$= 20$ N m

Total clockwise
moment about O
$= (8 \times 1) + (4 \times 3)$
$= 20$ N m

Total anticlockwise
moment about P
$= (10 \times 1) + (22 \times 1)$
$= 32$ N m

Total clockwise
moment about P
$= (8 \times 2) + (4 \times 4)$
$= 32$ N m

Equilibrium problem

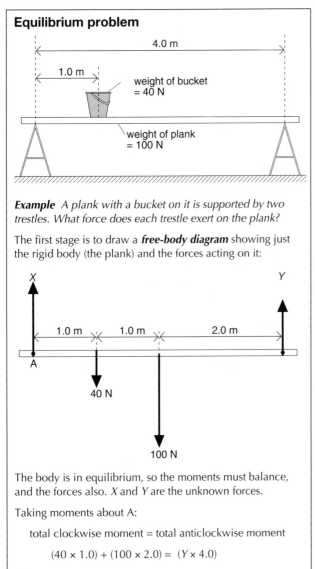

4.0 m

1.0 m

weight of bucket = 40 N

weight of plank = 100 N

Example A plank with a bucket on it is supported by two trestles. What force does each trestle exert on the plank?

The first stage is to draw a **free-body diagram** showing just the rigid body (the plank) and the forces acting on it:

X   Y

1.0 m   1.0 m   2.0 m

A

40 N

100 N

The body is in equilibrium, so the moments must balance, and the forces also. X and Y are the unknown forces.

Taking moments about A:

    total clockwise moment = total anticlockwise moment

$$(40 \times 1.0) + (100 \times 2.0) = (Y \times 4.0)$$

This gives $Y = 60$ N.

Note the advantage of taking moments about A: X has a zero moment, so it does not feature in the equation.

Comparing the vertical forces:

    total upward force = total downward force

$$Y + X = 40 + 100$$

As Y is 60 N, this gives $X = 80$ N.

The forces on the human forearm

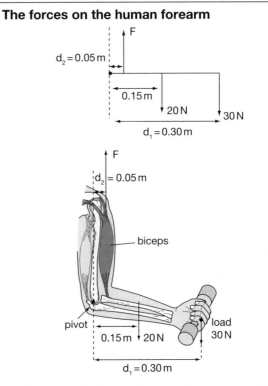

F

$d_2 = 0.05$ m

0.15 m

20 N

30 N

$d_1 = 0.30$ m

F

$d_2 = 0.05$ m

biceps

pivot

load 30 N

0.15 m   20 N

$d_1 = 0.30$ m

The forearm in the diagram has a weight of 20 N. To lift a weight of 30 N the biceps muscle exerts a force F.

Taking moments about the pivot point (the elbow):

$$Fd_2 = (20 \text{ N} \times 0.15 \text{ m}) + (30 \text{ N} \times d_1)$$

$$F \times 0.05 \text{ m} = 3.0 \text{ N m} + (30 \text{ N} \times 0.30 \text{ m}) = 12.0 \text{ N m}$$

$$F = 240 \text{ N}$$

Density

The density ρ of an object is calculated like this:

$$\rho = \frac{m}{V}$$

where m is mass in kg and V is volume in m^3

The SI unit of density is the kilogram/cubic metre (kg m^{-3}).

For example, 2000 kg of water occupies a volume of 2 m^3. So the density of water is 1000 kg m^{-3}.

Density values, in kg m^{-3}			
alcohol	800	iron	7 900
aluminium	2 700	lead	11 300

Pressure

Pressure p is calculated like this:

$$p = \frac{F}{A}$$

where F is force normal to area A in N and A is area in m^2

The SI unit of pressure is the newton/square metre, also called the **pascal** (**Pa**). For example, if a force of 12 N acts over an area of 3 m^2, the pressure is 4 Pa.

1.2.4 Car safety

Stopping distances

When a driver notices a hazard that requires an emergency stop, he or she has to react to the emergency and apply the brakes. The brakes then have to bring the car to rest.

stopping distance = thinking distance + braking distance

$$\text{thinking distance} = \frac{\text{speed of}}{\text{vehicle}} \times \frac{\text{driver's reaction}}{\text{time}} = vt$$

braking distance = distance travelled between brakes being applied and car stopping
(for a constant deceleration of the vehicle, a using $v^2 = u^2 + 2as$ gives $s = \frac{v^2}{2a}$)

The stopping distance d_m is given by

$$d_m = vt + \frac{v^2}{2a}$$

A typical reaction time is 0.7 s and a typical deceleration is 7.5 m s^{-2}.

When the speed of a car is doubled, its kinetic energy goes up by a factor of four. If the maximum braking force is applied, the braking distance will be four times as far at double the speed. The thinking distance will be doubled at double the speed.

The graph shows the stopping distances for this data.

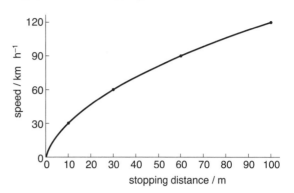

Stopping distance will increase if thinking distance is increased because the reaction time is increased, owing to driver fatigue or drugs including alcohol, for example. Braking distance will increase if:

- a car is heavily loaded

- the condition of the car's brakes is poor, so the braking force is less

- road conditions are poor, because of rain or ice for example, so friction is reduced.

Global positioning system

Satellite navigation systems use the Global Positioning System (GPS) to pinpoint the position of a vehicle within 10 m.

At any point on the Earth's surface there are four satellites in range. Each satellite has an accurate clock and broadcasts a radio signal with the current time. The receiver compares the signal broadcast time with the arrival time and uses the difference, and the speed of the signal, to work out the distance, d, to the satellite.

This places the receiver on a sphere with the satellite at the centre and a radius d.

The position is the found using trilateration. Using two satellites narrows the position to somewhere on the circle where the two spheres intersect. The third satellite gives a third sphere that intersects the circle at two points. The fourth satellite gives a fourth sphere that will go through only one of the points – and this is the position of the receiver.

Trilateration in two dimensions:

X is a from A
b from B
c from C

The radio signal travels at the speed of light, and the satellites orbit at 20 000 km, so the satellites have atomic clocks accurate to within a few nanoseconds.

Car safety features

In the event of a crash occurring, cars and equipment in them are designed to protect the occupants.

Large forces can kill or injure the occupants of a vehicle. There are several safety features that work on the principle that the force on a passenger is reduced by:

- reducing the velocity over a longer period of time, because $F = m\varnothing(v)/t$

- losing kinetic energy over a longer distance, because $F = \varnothing(E_K)/s$.

The design is such that the kinetic energy is reduced to zero in as long a time as possible but before the passenger hits a rigid part of the car.

The energy of the motion is converted into thermal energy and elastic energy.

Seat belts

When front seat occupants hit the windscreen, or rear seat passengers hit the front seats, they stop very quickly in a small distance and the force is large. When they wear seat belts, they move forward a small distance as the seat belt stretches slightly. This increases the distance (and time) during which they stop, so the force is reduced. After a crash the seat belt may have stretched beyond its elastic limit and should be replaced.

It is also better to be stopped by the large surface area of the seat-belt strap, than a hard part of the car with a small surface area.

Air bags

An air bag is a flexible nylon bag used for cushioning the impact with hard rigid parts of a vehicle, especially the steering wheel or dashboard. It slows the body and prevents it hitting a hard surface at high speed.

The steps in the deployment of the air bag are:

- a sensor (or sensors) detect the collision and produce an electric signal

- an initiator burns when it receives the electrical signal and ignites the inflator, which provides the gas

- the gas inflates the bag

- the bag gradually deflates.

Air bag sensors

An accelerometer is a sensor that detects sudden deceleration of the vehicle. A simple design is a mass on a spring. When the vehicle decelerates, the mass continues to move. The spring compression depends on the deceleration. When compressed enough, it switches on an electric circuit; for example, a magnetized mass can close a reed switch.

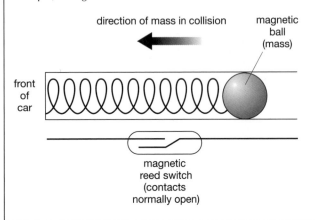

Much smaller, improved accelerometers are now used. For example, some MEMS (micro electromechanical systems) accelerometers use a change in electrical properties in part of an integrated circuit. Some designs use a small moving or tilting mass on part of the chip to change the separation of metal plates, which changes the capacitance of the circuit.

Other sensors include those that detect the wheel speed and those that detect whether the passenger seat is occupied, to avoid the air bag being deployed unnecessarily.

Air bag inflators

Older inflators contain sodium azide and potassium nitrate, which produce nitrogen gas. Sodium azide is toxic. Today other gas-generating chemicals are used, and there are also designs using compressed gas, and designs that use a mixture of both methods.

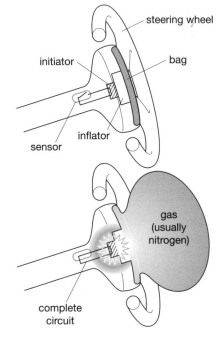

The bag is deployed and inflates in about 0.05 s. It must be fully inflated before the body hits it, otherwise it could do damage as it is moving at high speed towards the body. For this reason, it is important to wear a seat belt to slow the body down and give the air bag time to inflate. The air bag then deflates over the next 0.25 s, through vents in the side, bringing the body slowly to a stop.

Crumple zones

Cars are designed so that there is a rigid central safety cage for the occupants, surrounded by crumple zones at the front and rear of the car. These zones are designed to crumple on impact, so that kinetic energy is transformed by deforming the metal and heating it. While this happens, the occupants can continue to move forward, increasing the time and distance over which they stop and reducing the stopping force.

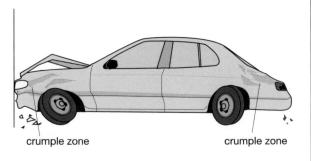

Module 3: Work and energy

1.3.1 Work and conservation of energy

Work

Work is done whenever a force makes something move. It is calculated like this:

work done = force × distance moved in direction of force

The SI unit of work is the **joule** (J). For example, if a force of 2 N moves something a distance of 3 m, then the work done is 6 J.

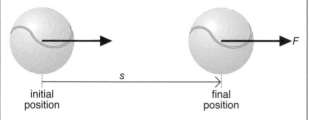

initial position   final position

Above, F is the resultant force on an object. If W is the work done when the force has caused a displacement s, then

$W = Fs$

When the direction of the force is not in the same direction as the displacement, the displacement in the direction of the force is $s \cos\theta$.

initial position

$s \cos \theta$

$W = F s \cos \theta$

Force-displacement graphs

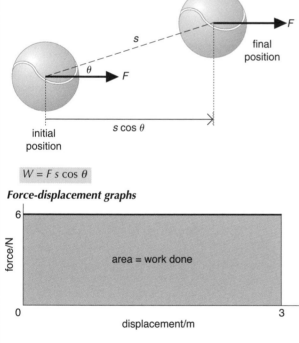

area = work done

Conservation of energy

According to the *law of conservation of energy*,

energy cannot be made or destroyed, but it can be changed from one form to another.

The diagram below shows the sequence of energy changes which occur when a ball is kicked along the ground. At every stage, energy is lost as heat. Even the sound waves heat the air as they die away. As in other energy chains, all the energy eventually becomes internal energy.

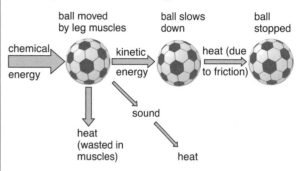

ball moved by leg muscles   ball slows down   ball stopped

chemical energy → kinetic energy → heat (due to friction)

sound

heat (wasted in muscles)   heat

Whenever there is an energy change, work is done – although this may not always be obvious. For example, when a car's brakes are applied, the car slows down and the brakes heat up, so kinetic energy is being changed into internal energy. Work is done because tiny forces are making the particles of the brake materials move faster.

An energy change is sometimes called an energy transfer. Whenever it takes place,

work done = energy transferred

So, for each 1 J of energy transferred, 1 J of work is done.

The graph above is for a uniform force of 6 N. When the displacement is 3 m, the work done is 18 J. Numerically, this is equal to the area under the graph between the 0 and 3 m points.

The area under a force-displacement graph is equal to the work done by the force. The following graph is for a non-uniform force, for example the force used to stretch a spring.

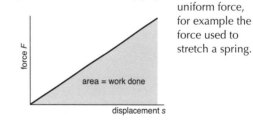

area = work done

displacement s

Energy

Things have energy if they can do work. The SI unit of energy is also the joule (J). You can think of energy as a 'bank balance' of work which can be done in the future.

When something does work, its energy decreases. When work is done on something, its energy increases.

A useful model of energy is to say that energy exists in different forms and can be transformed from one to another. In this model the forms are:

Kinetic energy This is energy which something has because it is moving.

Potential energy This is energy which something has because of its position, shape, or state. A stone about to fall from a cliff has *gravitational* potential energy. A stretched spring has *elastic* potential energy. Foods and fuels have *chemical* potential energy. Charge from a battery has *electrical* potential energy. Particles from the nucleus (centre) of an atom have *nuclear* potential energy.

Calculating work done and transfer of energy
Example

The resultant force stopping a car is 7.70 kN. The braking distance is 75.0 m. What was the kinetic energy of the car before braking?

Work done by stopping force = loss of kinetic energy

Fx = kinetic energy at the start – 0

Kinetic energy = 7.70×10^3 N $\times$ 75.0 m = 5.775×10^5 J = 5.78×10^5 J (3sf)

Internal energy Matter is made up of tiny particles (e.g. atoms or molecules) which are in random motion. They have kinetic energy because of their motion, and potential energy because of the forces of attraction trying to pull them together. An object's internal energy is the total kinetic and potential energy of its particles.

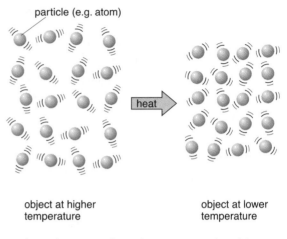

particle (e.g. atom)

object at higher temperature

object at lower temperature

Heat (thermal energy) This is the energy transferred from one object to another because of a temperature difference. Usually, when heat is transferred, one object loses internal energy, and the other gains it.

Radiant energy This is often in the form of waves. Sound and light are examples.

Note:
*   Kinetic energy, and gravitational and elastic potential energy are sometimes known as *mechanical energy*. They are the forms of energy most associated with machines and motion.
*   Gravitational potential energy is sometimes just called potential energy (or E_p), even though there are other forms of potential energy as described above.

1.3.2 Kinetic and potential energies

Calculating kinetic energy (E_k)

The stone below has kinetic energy. This is equal to the work done in increasing the velocity from zero to v.
The result is

> kinetic energy $E_k = \frac{1}{2}mv^2$

For example, if a 2.0 kg stone has a speed of 10 m s^{-1},
its $E_k = \frac{1}{2} \times 2.0 \times 10^2 = 100$ J

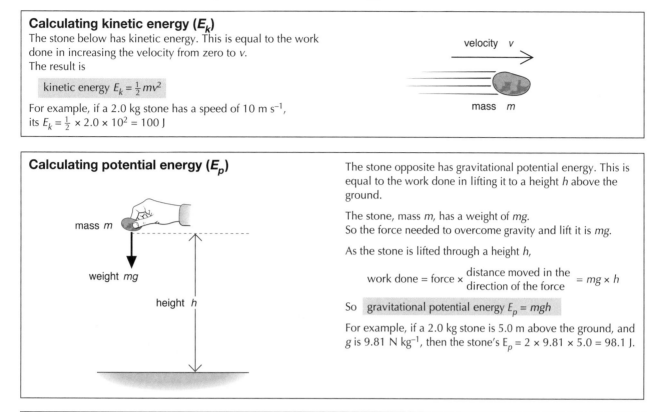

velocity v

mass m

Calculating potential energy (E_p)

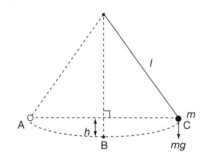

mass m

weight mg

height h

The stone opposite has gravitational potential energy. This is equal to the work done in lifting it to a height h above the ground.

The stone, mass m, has a weight of mg.
So the force needed to overcome gravity and lift it is mg.

As the stone is lifted through a height h,

$$\text{work done} = \text{force} \times \frac{\text{distance moved in the}}{\text{direction of the force}} = mg \times h$$

So   gravitational potential energy $E_p = mgh$

For example, if a 2.0 kg stone is 5.0 m above the ground, and g is 9.81 N kg^{-1}, then the stone's $E_p = 2 \times 9.81 \times 5.0 = 98.1$ J.

An exchange of E_p and E_k

A pendulum bob of mass m is released at point A and swings, stopping at point C and then returning to point A. The length of the string is l.

A○---- h ----B---- C m
      l mg

At points A and C the kinetic energy, $E_k = 0$ and the gravitational potential energy is a maximum.

At point B the kinetic energy is a maximum and the gravitational potential energy is a minimum.

If energy losses are negligible the pendulum will continue to swing to the same height each time, and $E_k + E_p$ is a constant value.

Between points B and C:

Loss in kinetic energy = gain in gravitational potential energy
$$\frac{1}{2}mv^2 = mgh$$

So the speed of the pendulum bob at point B, $v = \sqrt{(2gh)}$

The energy of a falling object

The diagram on the right shows how gravitational potential energy is changed into kinetic energy when something falls. The stone in this example starts with 100 J of gravitational potential energy. Air resistance is assumed to be zero, so no energy is lost to the air as the stone falls. Energy is conserved so

> E_p lost = E_k gained

By the time the stone is about to hit the ground (with velocity v), all of its potential energy has been changed into kinetic energy. So

$$\frac{1}{2}mv^2 = mgh$$

Dividing both sides by m and rearranging gives

> $v = \sqrt{2gh}$

In this example, $v = \sqrt{2 \times 9.81 \times 5.0} = 9.9$ m s^{-1}.

Note that v does not depend on m. A heavy stone hits the ground at exactly the same speed as a light one.

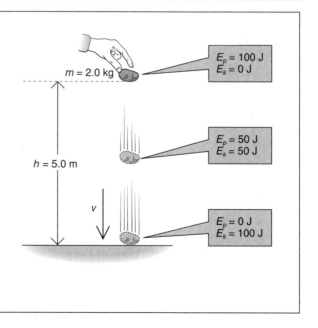

$m = 2.0$ kg

$E_p = 100$ J
$E_k = 0$ J

$h = 5.0$ m

$E_p = 50$ J
$E_k = 50$ J

v

$E_p = 0$ J
$E_k = 100$ J

1.3.3 Power

Power

> Power is the rate of work done.

It is calculated like this:

$$\text{power} = \frac{\text{energy transferred}}{\text{time taken}} \quad \text{or} \quad \text{power} = \frac{\text{work done}}{\text{time taken}}$$

The SI unit of power is the **watt** (W). A power of 1 W means that energy is being transformed at the rate of 1 joule/second ($J\,s^{-1}$), so work is being done at the rate of $1\,J\,s^{-1}$.

Calculating power
Example

To calculate the power output of an electric motor which raises a mass of 2.0 kg through a height of 12 m in 3.0 s:

$$\begin{aligned} E_p \text{ gained} &= mgh \\ &= 2.0 \times 9.81 \times 12 = 235.44\,J \end{aligned}$$

$$\text{power} = \frac{\text{energy transferred}}{\text{time taken}}$$

$$= \frac{235.44}{3.0} = 78\,W \text{ (2sf)}$$

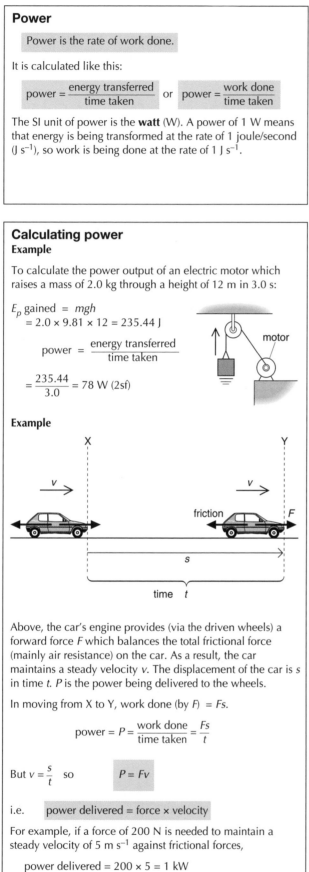

Example

Above, the car's engine provides (via the driven wheels) a forward force F which balances the total frictional force (mainly air resistance) on the car. As a result, the car maintains a steady velocity v. The displacement of the car is s in time t. P is the power being delivered to the wheels.

In moving from X to Y, work done (by F) $= Fs$.

$$\text{power} = P = \frac{\text{work done}}{\text{time taken}} = \frac{Fs}{t}$$

But $v = \dfrac{s}{t}$ so $\qquad \boxed{P = Fv}$

i.e. $\qquad$ power delivered = force × velocity

For example, if a force of 200 N is needed to maintain a steady velocity of 5 m s^{-1} against frictional forces,

$\qquad$ power delivered = 200 × 5 = 1 kW

All of this power is wasted as heat in overcoming friction. Without friction, no forward force would be needed to maintain a steady velocity, so no work would be done.

Efficiency

When a device, for example a motor, transfers energy some of the energy is always lost as heat.

The efficiency of a device is always less than 100% because of heat losses.

Efficiency is calculated like this:

$$\text{efficiency} = \frac{\text{useful energy output}}{\text{energy input}} = \frac{\text{useful power output}}{\text{power input}}$$

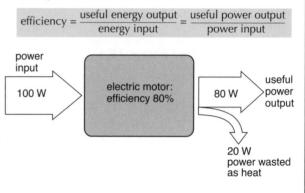

For example, if an electric motor's power input is 100 W, and its useful power output (mechanical) is 80 W, then its efficiency is 0.8. This can be expressed as 80%.

Sankey diagrams

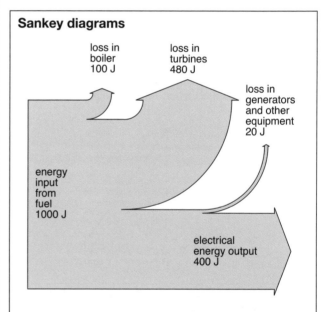

This diagram is a Sankey diagram for a power station. It shows what happens to every 1000 J of energy supplied as fuel to the power station. 400 J is converted to useful electrical energy but the rest is wasted as heat. The diagram is drawn to scale with the width of each arrow representing the amount of energy.

1.3.4 Behaviour of springs and materials

Deformation

The particles of a solid may be atoms, or molecules (groups of atoms), or ions. They are held closely together by electric forces of attraction.

When external forces are applied to a solid, its shape changes: *deformation* occurs. This alters the relative positions of its particles.

Forces on solids can be:

tension – the solid is stretched and you can measure the extension

compression – the solid is squashed and you can measure the compression.

Bending is a combination of tension and compressive strain.

There are two types of deformation, as described below.

Elastic deformation If the deformation is *elastic*, then the material returns to its original shape when the forces on it are removed.

Plastic deformation If the deformation is *plastic*, then the material does not return to its original shape when the forces on it are removed. For example, Plasticine takes on a new shape when stretched.

The force constant
Example

This graph shows the results of stretching a wire.

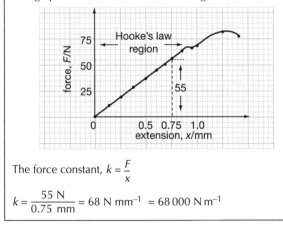

The force constant, $k = \dfrac{F}{x}$

$k = \dfrac{55\,\text{N}}{0.75\,\text{mm}} = 68\,\text{N}\,\text{mm}^{-1} = 68\,000\,\text{N}\,\text{m}^{-1}$

Hooke's law

If a material obeys **Hooke's law** then, for an elastic deformation the extension, x, (or the compression) is proportional to the force applied F.

$F \propto x$

$F = kx$ where k is the *force constant* of the spring or wire. It is the force per unit extension (or compression)

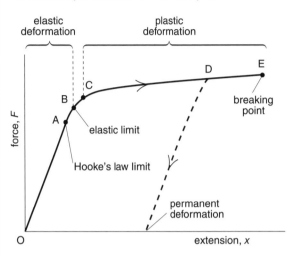

OA is the region where Hooke's law is obeyed.

B is the elastic limit.

OB is the region where the material will return to its original length when the force on it is removed.

C is the point beyond which the material deforms plastically – a small additional force produces a large extension.

This shape of graph is for a metal, other materials will have different graphs depending on their properties, for example some do not deform plastically, and some do not obey Hooke's law at all.

Stress and strain

On the right, a wire of cross-sectional area A is under tension from a force F (at each end). The *tensile stress* σ on the wire is defined like this:

$$\frac{\text{tensile}}{\text{stress}} = \frac{\text{force}}{\text{area}} \qquad \sigma = \frac{F}{A}$$

The unit of tensile stress is the pascal (Pa)
1 Pa = 1 Nm^{-2}.

The wire stretches so that its length L increases by x, called its **extension**. The *tensile strain* ε is defined like this:

$$\frac{\text{tensile}}{\text{strain}} = \frac{\text{extension}}{\text{original length}} \qquad \varepsilon = \frac{x}{L}$$

Tensile strain has no units.

Note:
- There are stresses and strains linked with compression and twisting. On these pages however, the word stress or strain by itself will imply the tensile type.

Work done and elastic potential energy

Stretching materials that obey Hooke's law

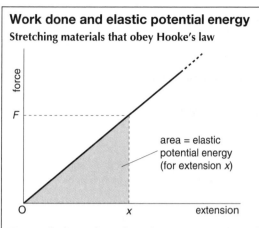

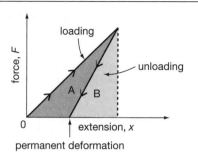

permanent deformation

The graph above shows how the extension varies with the stretching force for a material which obeys Hooke's law. The work done for an extension x is given by the shaded area. The area of a triangle $= \frac{1}{2} \times$ base $\times$ height. So the work done $= \frac{1}{2} Fx$.

As work is done *on* the material, energy is stored *by* the material. This is its ***elastic potential energy***. So

$$E = \frac{1}{2} Fx$$

using $F = kx$ where k is the force constant

$$E = \frac{1}{2} k x^2$$

When the force is removed the elastic potential energy is released.

The unloading line on the graph for a metal (or anything which obey's Hooke's Law) will be parallel to the loading line.

Stretching materials past the elastic limit:

The graph (top right) is for a wire that has been stretched past its elastic limit so that it has a permanent deformation. The work done stretching the wire is given by the areas $A + B$. When the force is removed the energy given by area B is recovered, this has been stored in the wire as elastic potential energy. The area A represents energy that has been transferred as heat; the wire will have heated up as it stretched.

Stretching materials that do not obey Hooke's law:

If Hooke's law is not obeyed, the work done is still equal to the area under a force-extension graph, but the equations given above do not apply because the area is not a triangle.

Stretching a polymer:

Materials like rubber are elastic and return to their original length, but when work is done to stretch them they do not store all the energy as potential energy, the long chain molecules of the polymer unravel, and the material heats up. The graph below shows the different loading and unloading curves for rubber.

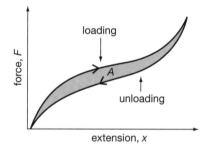

The area A in between the two lines is the energy lost as heat during the loading. (For more details see 'Stretching a polymeric material – rubber' page 44.)

An experiment to determine the Young modulus

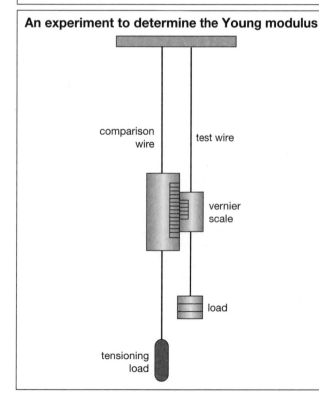

The wire to be tested is clamped and a hanger is attached.

A second wire is used for reference as this will cancel out errors due to changes in temperature, or bending of the support. The reference wire has a weight hung on it to ensure that it is kept straight and vertical.

Safety goggles must be worn because if the wire snaps the stored elastic potential energy could cause it to move suddenly and unpredictably. A box should be placed to safely catch the weights if the wire snaps.

Weights are added to the hanger and the extension of the wire is measured using the vernier scale. The diameter of the wire is measured using a micrometer. (See 'Reading a micrometer' page 12)

After each reading the weights should be removed to check that the elastic limit has not been exceeded.

To reduce errors the wire should be as long and thin as possible, so that the extension is larger and the percentage error in the extension is smaller.

Take care when clamping the wires to spread the clamping force over a large area because if the wires are tightly pinched they are likely to snap at these points when they are loaded.

Stress-strain graphs

This graph shows how stress varies with strain when a metal wire (steel) is stretched until it breaks. The force-extension graph in the section on Hooke's law will change according to the size and shape of the material. The advantage of the stress-strain graph is that it depends only on the material and not, for example, on its thickness or length.

Note:
- By convention, stress is plotted along the vertical axis and strain is plotted along the horizontal axis.

> If a material obeys **Hooke's law** then, for an *elastic* deformation, the strain is proportional to the stress.

The wire obeys Hooke's law up to point A.

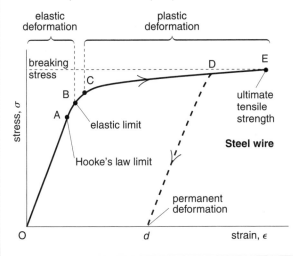

Mechanical properties

These describe how a material behaves when forces are applied. They depend on its structure, the strength of the bonds, and the type and number of defects present.

Strength A *strong* material has a high ultimate tensile stress i.e. a high stress is needed to break it.

Ductility A *ductile* material can be drawn into wires.

Malleability A *malleable* material can be hammered into different shapes.

Stiffness A *stiff* material has a high Young modulus, i.e. a high stress produces little strain.

Toughness A *tough* material will deform plastically before it breaks. It is not brittle, i.e. cracks do not easily spread. It can withstand dynamic loads such as shock or impact.

Hardness A *hard* material cannot be easily scratched or indented. There are no absolute values for hardness. One material is harder than another if it will scratch it.

Durability A *durable* material can be repeatedly loaded and unloaded without its properties deteriorating.

Brittleness A *brittle* material cracks and breaks without plastic deformation.

Smoothness A *smooth* material has a low friction surface. This means you can treat friction as negligible.

O to B The deformation of the wire is elastic.

B This is the **elastic limit**. Beyond it, the deformation becomes plastic as layers of particles slide over each other. If the stress were removed at, say, point D, the wire would be left with a permanent deformation (strain *d* on the axis).

C This is the **point** beyond which little extra force is needed to produce a large extra extension. If a material can be stretched like this, it is said to be **ductile**.

E The wire develops a thin 'neck', then a **ductile fracture** occurs. The highest stress just before the wire breaks is called the **ultimate tensile strength** or the breaking stress.

Fatigue If a metal is taken through many cycles of *changing* stress, a fatigue fracture may occur before the ultimate tensile stress is reached. Fatigue fractures are caused by the slow spread of small cracks.

Creep This is the deformation which goes on happening in some materials if stress is maintained. For example, unsupported lead slowly sags under its own weight.

Stretching a ductile material – steel

When metals are stretched they produce a stress strain graph similar to the graph for steel wire (see 'Stress-strain graphs' box above). Metals are ductile and malleable.

Stretching a brittle materials – glass

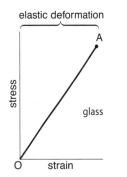

The graph above shows what happens if increasing tensile stress is applied to a glass thread. Elastic deformation occurs until, at point A, a crack suddenly grows, and the glass breaks. A material which behaves like this is said to be **brittle**. The break is called a **brittle fracture**.

Stretching a polymeric material – rubber

Polymers have long-chain molecules, each of which may contain many thousands of atoms. In a polymer, the chains may be coiled up and tangled like spaghetti.

Rubber and wool are natural polymers. Plastics, such as nylon and artificial rubber, are synthetic polymers.

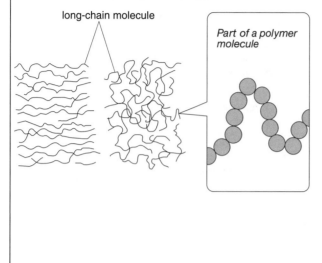

long-chain molecule

Part of a polymer molecule

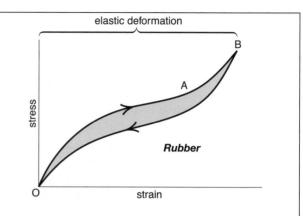

elastic deformation

stress

strain

Rubber

The graph shows what happens if increasing stress is applied to a rubber cord, and then released before the breaking point. Rubber does not obey Hooke's law. Also, much higher strains are possible than in steel or glass. For example, if the extension is twice the original length, the strain is 2.

O to A The molecular chains in the rubber are being uncoiled and straightened.

A to B The chains are almost straight, so the rubber is becoming proportionately more difficult to stretch.

B to A The rubber contracts when the stress is removed.

During this cycle of extension and contraction, energy is lost as heat. The effect is called *elastic hysteresis*. The shaded area represents the energy lost per unit volume.

Unit G482 Electrons, waves and photons

Module 1: Electric current

2.1.1 Charge and current

Electric charge and electric current

There are two types of **electric charge**, positive and negative. When charged particles are moving there may be a net flow of charged particles in one direction. This is called an **electric current**.

In a metal the current is due to the movement of **electrons**.

An atom that loses or gains orbital electrons is called an **ion**. Negatively charged ions are atoms that have gained electrons and positively charged ions are atoms that have lost electrons. An **electrolyte** is a substance that contains ions which are free to move, so electric current can pass through an electrolyte due to the movement of ions.

Conventional current and electron flow

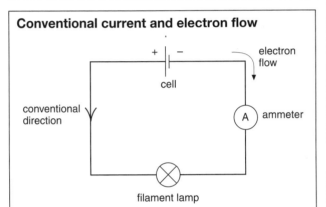

In the circuit above, chemical reactions in the cell push electrons out of the negative (–) terminal, round the circuit, to the positive (+) terminal. This net flow of electrons forms an **electric current**.

An arrow in the circuit indicates the direction from the + terminal round to the –. Called the **conventional direction**, it is the *opposite* direction to the actual electron flow.

The SI unit of current is the **ampere** (A).

Current may be measured using an **ammeter** as above.

Kirchhoff's first law

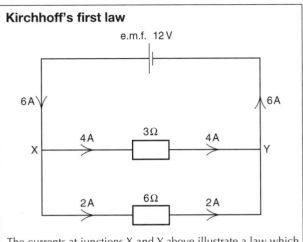

The currents at junctions X and Y above illustrate a law which applies to all circuits:

> total current out of junction = total current into junction

This is known as **Kirchhoff's first law**. It arises because, in a complete circuit, charge is never gained or lost. It is conserved. So the total rate of flow of charge is constant.

Electric charge

Charge ΔQ passing in time Δt gives an electric current I, where:

$$\Delta Q = I\Delta t$$

The SI unit of charge is the **coulomb** (**C**).

For example, if there is a current of 1 A for 1 s, the charge flowing is 1 C. (This is how the coulomb is defined.) Similarly, if there is a current of 2 A for 3 s, the charge flowing is 6 C.

An electron has a negative charge of $e = 1.6 \times 10^{-19}$C. This is called the elementary charge.

Current and mean drift velocity

Most electrons are bound to their atoms. However, in a metal, some are **free electrons** which can move between atoms. When a p.d. is applied, and a charge flows causing a current, the free electrons are the **charge carriers**.

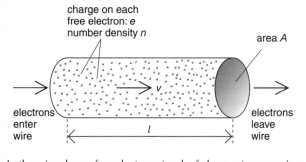

charge on each
free electron: e
number density n

area A

electrons enter wire

v

l

electrons leave wire

In the wire above, free electrons (each of charge e) are moving with an average velocity v. n is the **number density** of free electrons: the number per unit volume (per m³).

In the wire the number of free electrons = Anl
So total charge carried by free electrons = $Anle$

As time = distance ÷ velocity:
time taken for all the free electrons to pass through A

$$= \frac{l}{v}$$

As current I = charge/time

$$I = \frac{nAle \times v}{l}$$

$$\therefore \quad \boxed{I = Anev}$$

v is called the mean **drift velocity**. Typically, it can be less than a millimetre per second for the current in a wire.

- The number density of free electrons is different for different metals. For copper, it is 8×10^{28} m⁻³.
- When liquids conduct, ions are the charge carriers. The above equations apply, except that e and n must be replaced by the charge and number density of the ions.
- Conductors, for example metals, have high number densities, n.
- Insulators have very low number densities. For a good insulator n is close to, or equal to, zero.
- Semiconductors have number densities much smaller than metals. Silicon is 4×10^{15} m⁻³ and germanium 3×10^{19} m⁻³. The number density can be altered by adding small amounts of impurity to the semiconductor. This is called doping.

Module 2: Resistance

2.2.1 Circuit symbols

Circuit symbols

These are standard component symbols

Conductors

—→— current direction, energy
or signal flow

—┼— conductors crossing with
no connection

—┬— junction of conductors
(dot recommended)

—┼┬— double junction of conductors
(dots recommended)

⏚ earth or ground

╲ switch

Cells

—╢├— primary or secondary cell

—╢╢╢├— battery of cells, form 1

—╢┆├— battery of cells, form 2 –
note nominal voltage

—○ ○— open terminals

Other components

⊟ relay coil

⊗ indicator or light source

—☐— fuse

—▭— heater

—▭— fixed resistor

—▱— potentiometer

⌣⌣⌣ transformer with magnetic core

(A) ammeter

(V) voltmeter

⊓ earphone

⊏⌷ loudspeaker

⌓ electric bell

▽ buzzer

(M) motor

[G] generator

—▷|— diode

—▷|— light-emitting diode

↘↘—|— photovoltaic cell

↘↘—▭— light dependent resistor

↘↘—▷|— photodiode

Components with variable values

⟋ placed across the component
symbol indicates a variability of
value which may, or may not,
be continuous, e.g.:

—▱— variable resistor

⟋ placed across the component
symbol indicates that it has an
inherent non-linear variability, e.g.:

—▱— thermistor

When drawing circuits make sure that:

- There are no gaps between components and wires.
- The symbols are accurately drawn, for example, 'A' in an ammeter, not 'a'.

When adding a voltmeter or ammeter to a circuit, remember:

Voltage across – connect in parallel with the component.

Current through – connect in series with the component.

2.2.2 E.m.f. and p.d.

E.m.f. and p.d.

An electric circuit supplies electrical energy to the components in the circuit. The **electromotive force (e.m.f.)** is a measure of the energy supplied to the electric charge from the energy source, for example from a battery, power supply, or photovoltaic cell. Note that this is a measure of energy and is not a force. (The name survives from a time when electricity was not well understood.)

> The **electromotive force (e.m.f.)** is the energy transferred per unit charge when energy is transferred *to* the electric charge.

When electric charge is in coulombs (C) and energy is in joules (J) the e.m.f. is in **volts (V)**

> 1 volt (V) = 1 joule of energy per coulomb of charge

When electric charge flows through a component some electrical energy is transferred from the charge to the component. The difference in energy per unit charge between when it enters and leaves the component is called the **potential difference (p.d.)**

> The **potential difference (p.d.)** is the energy transferred per unit charge when energy is transferred *from* the electric charge.

Electrical energy cannot be stored, so all of the electrical energy is transferred from the electric charge to the wires and components of a circuit. This means that, for example in the circuit below, the e.m.f. of the battery is numerically equal to the total p.d. across the circuit.

In the circuit below, several cells have been linked in a line to form a *battery*. The **e.m.f.** across the battery terminals is 12 volts (V).

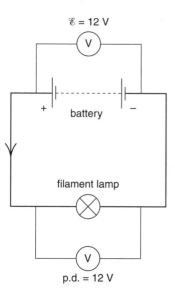

Connecting wires are treated as having negligible resistance so the p.d. across the bulb is also 12 V. This means that, for each coulomb pushed through it, 12 J of electrical energy is changed into other forms (heat and light energy).

p.d. may be measured using a **voltmeter** across a component in parallel as shown above.

p.d., energy, and charge are linked by this equation:

> energy transferred = p.d. × charge
> $W = VQ$

For example, if a charge of 2 C moves through a p.d. of 3 V, the energy transferred is 6 J.

When a battery is supplying current, some energy is wasted inside it, which reduces the p.d. across its terminals. For example, when a torch battery of e.m.f. 3.0 V is supplying current, the p.d. across its terminals might be only 2.5 V.

2.2.3 Resistance

Resistance

The resistance of a component is defined by the equation:

$$\text{resistance} = \frac{\text{potential difference}}{\text{electric current}}$$

$$R = \frac{V}{I}$$

The unit of resistance is the **ohm (Ω)**

There is 1 ohm (Ω) of resistance when a p.d of 1 volt causes a current of 1 ampere

$$1\ \Omega = 1\ \text{V A}^{-1}$$

Example

If a p.d. of 12 V causes a current of 40 mA the resistance is $(12\ \text{V} \div 40 \times 10^{-3}\ \text{A}) = 300\ \Omega$

Ohm's law

If a conductor obeys **Ohm's law**, then the current I through it is directly proportional to the p.d. V across it, provided the temperature is constant.

Metals obey Ohm's law. If a graph of I against V is plotted for a metal conductor at constant temperature, the result is as on the right. Expressed mathematically this is

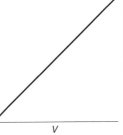

$$\frac{V}{I} = \text{constant}$$

Note: Resistance is always calculated using $R = V \div I$ so this equation alone is not an expression of Ohm's law. Ohm's law is that $R = V \div I$ is a constant value for all values of V and I. This is only true for metals if their temperature is kept constant.

I-V characteristics

The link between I and V can be investigated using the circuit opposite. Graphs for three different components are shown. These graphs of current I against p.d. V are called $I - V$ characteristics. To keep the temperature constant the circuit should be switched off between readings and, if necessary, time for it to cool allowed. Check for zero errors of the meters, and it is best to plot your graph as you do the experiment so that anomalous readings can be rechecked. (A negative V means that the DC supply connections have been reversed.)

variable
DC supply

component
under test

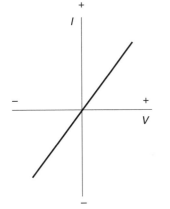

Wire (metal) kept at constant temperature V/I is constant, because R is constant.

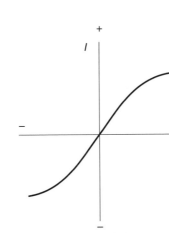

Lamp filament (metal) As the current rises the filament heats up. V/I increases, because R increases with temperature.

Diode (Light Emitting LED) R is extremely high in one direction. It is much lower in the other direction and decreases as the current rises. In effect, a diode allows charge to flow in one direction only. This is called the forward bias direction.

reverse bias   forward bias

Uses and benefits of using LEDs

Light emitting diodes are semiconductor devices. They are diodes that emit light when current flows through them. In recent years much brighter and white diodes have been developed and they have moved from being used as indicator lights, for example to show if equipment is on or off, to being used for illumination. Advantages are:

- They use low p.d.s.
- They have very long working life (100 000 hours compared to, for example, 5000 hours for a filament lamp.)
- They use low power compared to filament lamps, and save energy reducing running costs.

- They are very robust.
- They switch on instantly (so they can be used in communications technology).
- They are useful as a cluster of lights because if one blows the other still operate.

As the cost is reduced these advantages mean they are likely to become very widely used. They are already used for traffic lights, outdoor signs, flashlights and headlamps. They are used for alphanumeric displays, but are less popular now that better LCDs are available. Infrared LEDs are used in remote controls and communications.

2.2.4 Resistivity

Resistivity

The resistance of a metal conductor (such as a wire) depends on various factors:

- **Length** A long wire has more resistance than a short one.
- **Cross-sectional area** A thin wire has more resistance than a thick one.
- **Temperature** A hot wire has more resistance than a cold one.
- **Type of material** A nichrome wire has more resistance than a copper wire of the same dimensions.

So the resistance R of a conductor depends on its length l and cross-sectional area A:

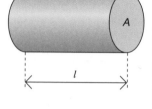

$$R \propto \frac{l}{A}$$

This can be changed into an equation by means of a constant, ρ, known as the **resistivity** of the material:

$$R = \frac{\rho l}{A}$$

> The resistivity (ρ) of a material is numerically equal to the resistance in ohms of a sample 1 metre long with cross sectional area 1 metre squared.

It can be calculated from $\rho = \dfrac{RA}{l}$

The unit of resistivity is ohm metre (Ω m)

(note this is ohms multiplied by metres, not ohms per metre).

Resistivities, in Ω m

copper: 1.55×10^{-8} aluminium: 2.50×10^{-8}

All materials, including insulators, can have resistivity. Resistivities range from 10^{-8} to 10^{14}.

Superconductivity

When some metals are cooled towards absolute zero, a **transition temperature** is reached at which the resistance suddenly falls to zero. This effect is called **superconductivity**. It occurs when there is no interaction between the free electrons and the lattice, and is explained by the quantum theory. Some specially developed metal compounds have transition temperatures above 100 K.

If an electromagnet has a superconducting coil, a huge current can be maintained in it with no loss of energy. This enables a very strong magnetic field to be produced.

Resistance and temperature

A conducting solid is made up of a **lattice** of atoms. When a current flows, electrons move through this lattice.

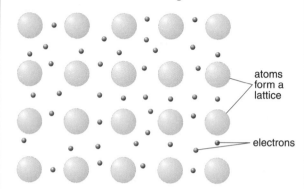

atoms form a lattice

electrons

Metals When free electrons drift through a metal, they make occasional collisions with the lattice. These collisions are inelastic and transfer energy to the lattice as internal energy. That is why a metal has resistance. If the temperature of a metal rises, the atoms of the lattice vibrate more vigorously. Free electrons collide with the lattice more frequently, which *increases the resistance*.

Semiconductors (e.g. silicon) At low temperature, the electrons are tightly bound to their atoms. But as the temperature rises, more and more electrons break free and can take part in conduction. This easily outweighs the effects of more vigorous lattice vibrations, so the *resistance decreases*. At around 100–150 °C, **breakdown** occurs. There is a sudden fall in resistance – and a huge increase in current. That is why semiconductor devices are easily damaged if they start to overheat.

Semiconductor devices include diodes, light dependent resistors (LDRs) and thermistors. A thermistor has a resistance which changes with temperature. The most common thermistors are negative temperature coefficient (NTC) thermistors, which means that their resistance decreases with increasing temperature. This graph is for a typical metal conductor and an NTC thermistor.

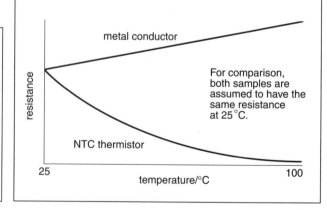

metal conductor

resistance

For comparison, both samples are assumed to have the same resistance at 25 °C.

NTC thermistor

25   temperature/°C   100

2.2.5 Power

Energy transfer and power

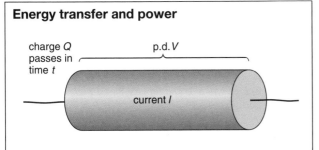

charge Q passes in time t

p.d. V

current I

Above, charge Q passes through a resistor in time t. Work W is done by the charge, so energy W is transferred – the electrons lose electrical potential energy and the lattice gains internal energy (it heats up).

W, Q, and V are linked by this equation (see also E1):

$$W = QV$$

But $Q = It$, so $\quad \boxed{W = IVt} \quad$ (1)

Applying $V = IR$ to the above equation gives

$$\boxed{W = I^2Rt} \quad \text{and} \quad \boxed{W = V^2t/R} \quad (2)$$

For example, if a current of 2 A flows through a 3 Ω resistor for 5 s, $W = 2^2 \times 3 \times 5 = 60$ J. So the energy dissipated is 60 J. *Double* the current gives *four* times the energy dissipation.

Note:
- Equation (1) can be used to calculate the total energy transferred whenever electrical potential energy is changed into other forms (e.g. rotational kinetic energy and internal energy in an electric motor). Equations (2) are only valid where *all* the energy is changed into internal energy. Similar comments apply to the power equations which follow.

Power is the rate of energy transfer

$$\text{power} = \frac{\text{energy transfer}}{\text{time}}$$

As power $P = W/t$, it follows from (1) and (2) that

$$\boxed{P = VI} \qquad \boxed{P = I^2R} \qquad \boxed{P = \frac{V^2}{R}}$$

The cost of energy
A domestic electricity meter records the number of units of electricity consumed. The unit is a kilowatt hour. The price of a unit is set by the power company.

Example

A unit of electricity costs 15p. A 3 kW fan heater is used for 2 hours every day. How much does it cost per week?

Time used in a week = $2 \times 7 = 14$ hours

Energy used (kW h) = power (kW) x time(h) = 3 kW $\times$ 14 h = 42 kW h

Cost of electricity = units of energy used (kW h) x cost of one unit (p per kW h)

Cost = 42 kW h $\times$ 15p per kW h

Cost = 630p = £6.30

Fuses
A fuse is a safety device in a circuit that can prevent a circuit from overheating.

When there is an electric current in a circuit the wires and components heat up. If too large a charge flows this can damage or melt parts of the circuit or cause a fire. A fuse melts first, before other wires or components are damaged or catch fire. This breaks the circuit and stops the current. The fuse is a piece of wire that is thinner than the rest of the connecting wires, or made of a metal with a lower melting point, or both.

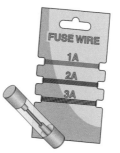

A fuse is often a thin wire contained inside a cartridge as shown in the diagram. Sometimes it is just a piece of wire.

A fuse is also used to prevent metal appliances becoming 'live'. If the live wire touched the earthed metal case so that the appliance was at a voltage of 230 V, a large current would flow through the fuse to earth. This would melt the fuse and cut off the electricity supply.

Note: A fuse is a safety device, but it is important to realise that, while a fuse can prevent a fire from starting, or a metal appliance becoming live, if you touch a live wire it will not prevent you receiving an electric shock. A residual current circuit breaker is needed to cut the current off very quickly.

Choosing the correct fuse
The current at which the fuse melts must be larger than the normal operating current of the circuit, but not so much larger that it does not melt when there is a fault.

Example

A 2.70 kW kettle has a normal operating current
$I = 2.70 \times 10^3$ W $\div$ 230 V = 11.7 A

Fuses for 3 pin plugs are usually 3 A or 13 A, so a 13 A fuse is suitable.

A 100 W table lamp has a normal operating current I = 100 W $\div$ 230 V = 0.43 A so a 3 A fuse is suitable.

Energy (in kW h) and power
For practical reasons, energy is sometimes measured in kilowatt hours (kW h) instead of joule (J).

1 *kilowatt hour* (kW h) is the energy supplied when delivered at the rate of 1 kW (i.e. 1000 J s^{-1}) for 1 hour.

$$\text{energy} = \text{power} \times \text{time}$$

So 1 kW h = 1000 J s^{-1} $\times$ 3600 s = 3.6×10^6 J

Module 3: DC circuits

2.3.1 Series and parallel circuits

Kirchhoff's second law

This law is a consequence of the conservation of energy. The sum of the e.m.f.s is the total electrical energy supplied to the charge flowing in the circuit and the sum of the p.d.s is the total energy transferred from the charge flowing in the circuit.

The charge in the circuit is conserved, so because of the conservation of energy the sum of the e.m.f.s must be equal to the sum of the p.d.s.

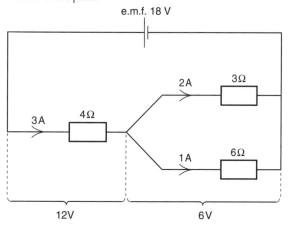

The arrangement above is called 'a circuit'. But, really, there are *two* complete circuits through the battery. To avoid confusion, these will be called *loops*.

In the circuit above, charge leaves the battery with electrical potential energy. As the charge flows round a loop, its energy is 'spent' – in stages – as heat. The principle that the total energy supplied is equal to the total energy spent is expressed by **Kirchhoff's second law**.

> Round any closed loop of a circuit, the algebraic sum of the e.m.f.s is equal to the algebraic sum of the p.d.s (i.e. the algebraic sum of all the IRs).

Note:
- From the law, it follows that if sections of a circuit are in parallel, they have the same p.d. across them.
- 'Algebraic' implies that the direction of the voltage must be considered. For example, in the circuit below, the e.m.f. of the right-hand battery is taken as *negative* (–4 V) because it is opposing the current. Therefore:

algebraic sum of e.m.f.s $= 18 + (-4) = +14$ V
algebraic sum of IRs $= (2 \times 3) + (2 \times 4) = +14$ V

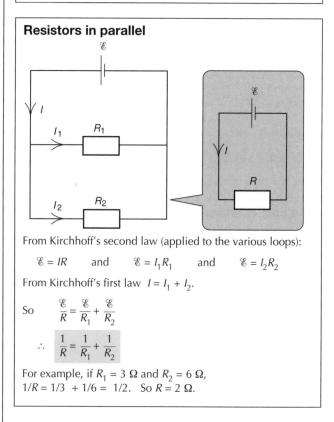

Resistors in series

If R_1 and R_2 below have a total resistance of R, then R is the single resistance which could replace them.

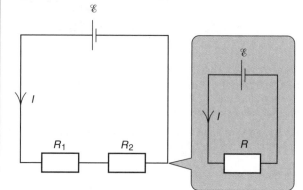

From Kirchhoff's first law, all parts of the circuit have the same current I through them.

From Kirchhoff's second law $\mathscr{E} = IR$ and $\mathscr{E} = IR_1 + IR_2$.

So $\quad IR = IR_1 + IR_2$

$\therefore \quad \boxed{R = R_1 + R_2}$

For example, if $R_1 = 3\ \Omega$ and $R_2 = 6\ \Omega$, then $R = 9\ \Omega$.

Resistors in parallel

From Kirchhoff's second law (applied to the various loops):

$\mathscr{E} = IR \quad$ and $\quad \mathscr{E} = I_1 R_1 \quad$ and $\quad \mathscr{E} = I_2 R_2$

From Kirchhoff's first law $I = I_1 + I_2$.

So $\quad \dfrac{\mathscr{E}}{R} = \dfrac{\mathscr{E}}{R_1} + \dfrac{\mathscr{E}}{R_2}$

$\therefore \quad \boxed{\dfrac{1}{R} = \dfrac{1}{R_1} + \dfrac{1}{R_2}}$

For example, if $R_1 = 3\ \Omega$ and $R_2 = 6\ \Omega$,
$1/R = 1/3 + 1/6 = 1/2$. So $R = 2\ \Omega$.

Circuit problems

Example

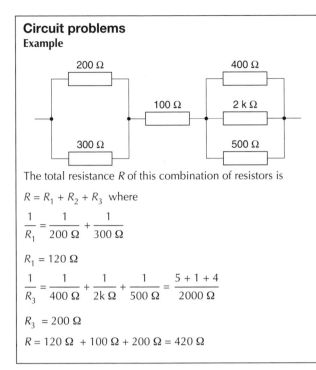

The total resistance R of this combination of resistors is

$R = R_1 + R_2 + R_3$ where

$$\frac{1}{R_1} = \frac{1}{200\ \Omega} + \frac{1}{300\ \Omega}$$

$R_1 = 120\ \Omega$

$$\frac{1}{R_3} = \frac{1}{400\ \Omega} + \frac{1}{2k\ \Omega} + \frac{1}{500\ \Omega} = \frac{5 + 1 + 4}{2000\ \Omega}$$

$R_3 = 200\ \Omega$

$R = 120\ \Omega + 100\ \Omega + 200\ \Omega = 420\ \Omega$

Using equations for e.m.f.

Example

A 'flat' car battery with an internal resistance of 0.05 Ω is charged using a battery charger with an e.m.f. of 15 V and internal resistance of 0.65 Ω

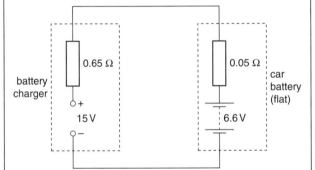

The total resistance of the circuit = 0.65 Ω + 0.05 Ω = 0.7 Ω

The sum of the e.m.f.s of the circuit = 15 V – 6.6 V = 8.4 V

Using $\mathscr{E} = I(R + r)$ with $R = 0$

The initial charging current = 8.4 V ÷ 0.7 Ω = 1.2 A

Example

A battery of e.m.f. 13 V is connected to a resistor with resistance of 6 Ω. The terminal p.d. is 12 V. What is the internal resistance of the battery?

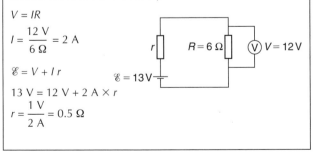

$V = IR$

$I = \dfrac{12\ \text{V}}{6\ \Omega} = 2\ \text{A}$

$\mathscr{E} = V + Ir$

$13\ \text{V} = 12\ \text{V} + 2\ \text{A} \times r$

$r = \dfrac{1\ \text{V}}{2\ \text{A}} = 0.5\ \Omega$

Internal resistance

On the opposite page, it was assumed that each battery's output p.d (the p.d across its terminals) was equal to its e.m.f. In reality, when a battery is supplying current, its output p.d. is *less* than its e.m.f. The greater the current, the lower the output p.d. This reduced voltage is due to energy dissipation in the battery. In effect, the battery has ***internal resistance***. Mathematically, this can be treated as an additional resistor in the circuit.

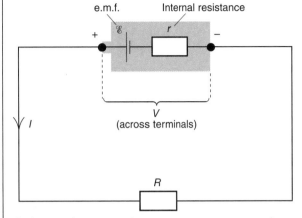

The battery above is supplying a current I to an external circuit. The battery has a constant internal resistance r.

From Kirchhoff's second law $\mathscr{E} = IR + Ir$

$\mathscr{E} = I(R + I)$ (1)

But $V = IR$, so $\mathscr{E} = V + Ir$

So $V = \mathscr{E} - Ir$ (2)

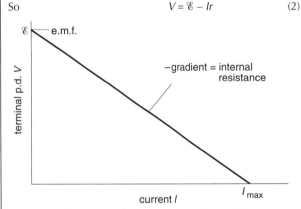

The graph above shows how V varies with I. Unlike earlier graphs, V is on the vertical axis.

Note:
- When I is zero, $V = \mathscr{E}$. In other words, when a battery is in ***open circuit*** (no external circuit), the p.d. across its terminals is equal to its e.m.f.
- When R is zero, V is zero. In other words, when the battery is in ***short circuit*** (its terminals directly connected), its terminal p.d. is zero. In this situation, the battery is delivering the maximum possible current, I_{max}, which is equal to $\mathscr{E}/r$. Also, the battery's entire energy output is being wasted internally as heat.
- As $I_{max} = \mathscr{E}/r$, it follows that $r = \mathscr{E}/I_{max}$. So the gradient of the graph is numerically equal to the internal resistance of the battery.

If both sides of equation (2) are multiplied by I, the result is $VI = \mathscr{E}I - I^2 r$. Rearranged, this gives the following:

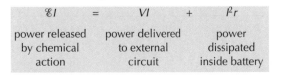

$\mathscr{E}I$	=	VI	+	$I^2 r$
power released by chemical action		power delivered to external circuit		power dissipated inside battery

2.3.2 Practical circuits

Potential divider

A *potential divider* or *potentiometer* like the one below passes on a fraction of the p.d. supplied to it.

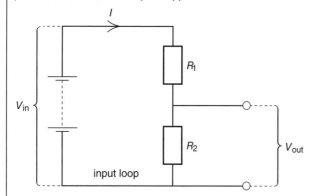

In the input loop above, the total resistance $= R_1 + R_2$.

So

$$I = V_{in}/(R_1 + R_2)$$

But $V_{out} = IR_2$, so $V_{out} = \left(\dfrac{R_2}{R_1 + R_2}\right)V_{in}$

For example, if R_1 and R_2 are both 2 kΩ, then $R_2/(R_1 + R_2)$ works out at 1/2, so V_{out} is a half of V_{in}.

A potential divider circuit can be used to produce a variable p.d. by varying one of the resistors R_1 or R_2 . This can be done by using a variable resistor or by using a sensor that has a resistance which changes with some other property, examples are:

- An NTC thermistor—resistance decreases as temperature increases (see Resistance and temperature, page 52)
- A **light-dependent resistor (LDR)** – resistance decreases with increased light level.

Example

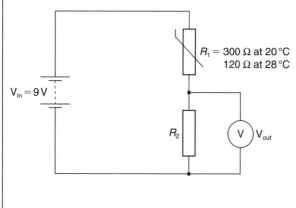

As the temperature increases the resistance of the thermistor, R_1 decreases. The p.d. across the thermistor also decreases so the output p.d., V_{out}, across R_2 increases. This output p.d. could be used to switch on a fan.

If the fan switches on when the p.d. reaches 5V, then to switch on the fan when the temperature reaches 28 °C:

$$5\ V = \frac{R_2}{(120\ \Omega + R_2)} \times 9\ V$$

$$5\ V \times 120\ \Omega + 5\ V \times R_2 = 9\ V \times R_2$$

$$R_2 = (600 \div 4)\ \Omega = 150\ \Omega$$

At 20°C the output voltage will be:

$$V_{out} = \frac{150\ \Omega}{(300\ \Omega + 150\ \Omega)} \times 9\ V \ = 3\ V$$

Example

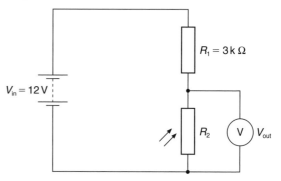

As the light level decreases the resistance of the LDR, R_2 increases. The p.d. across the LDR also increases so the output p.d., V_{out}, across R_2 increases. This output p.d. could be used to switch on a light.

If the LDR has a resistance of 9k Ω at twilight, the output p.d. at twilight is:

$$V_{out} = \frac{9\ \Omega}{(3k\ \Omega + 9k\ \Omega)} \times 12\ V = 9\ V$$

Notes:

- The above analysis assumes that no external circuit is connected across R_2. If such a circuit is connected, then the output p.d. is reduced.
- These two examples are sensor circuits which will activate electronic switches for power circuits containing devices such as fans or lamps.
- Potential dividers are not really suitable for high-power applications because of energy dissipation in the resistors.

Data-logging

A data-logger is an electronic device for recording data. The advantages of using data-logging rather than recording data manually are:

- Measurements are taken at the correct time. Unlike a human the computer will not forget to take a reading or take a reading too early or late.
- The system can be set to take measurements at some time in the future without a person having to be in attendance.

- Regular, frequent readings can be taken without a person needing a break.
- Mistakes are not made in reading or recording the results. Humans sometimes make mistakes, for example when reading a value off a scale.
- Data-logging devices can be sent to places that are dangerous or impossible for humans to go e.g. to Mars, into volcanoes, or onto a roof during a hurricane.
- Tables of results and graphs can be produced automatically.

Module 4: Waves

2.4.1 Wave motion

Types of wave motion

Waves transfer energy from one place to another. Where ever there is wave motion, there must be:

- a source of oscillation
- a material or field which can transmit oscillations.

Wave motion can be demonstrated using a 'slinky' spring, as shown below. The moving waves are called **progressive waves**. There are two main types.

Transverse waves The oscillations are at right-angles to the direction of travel:

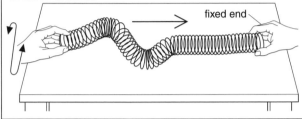

Longitudinal waves The oscillations are in line with the direction of travel, so that a compression ('squash') is followed by a rarefaction ('stretch'), and so on.

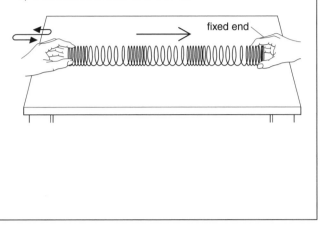

Progressive wave motion

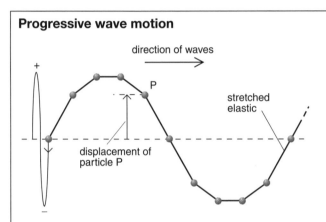

Above is one model of how waves travel. The first particle is oscillated up and down. This pulls on the next particle, making it oscillate up and down slightly later, and so on, along the line. As a result, **progressive** (moving) waves are seen travelling from left to right. The waves in this case are **transverse** because the oscillations are at right-angles to the direction of travel.

The **displacement** of a particle at any instant is measured from the centre line, with a + or – to indicate an *upward* or *downward* direction.

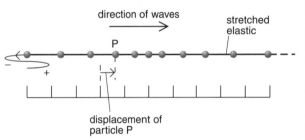

With **longitudinal** waves as above, the oscillations are in the direction of travel. So a particle can have a displacement to the *right* (+) or *left* (–).

The speed of waves depends on the properties of the **medium** (material) through which they are travelling. For example, if the particles above are lighter, or the elastic tighter, then each particle is affected more rapidly by the one before, so the wave speed is greater.

Wave features – key definitions

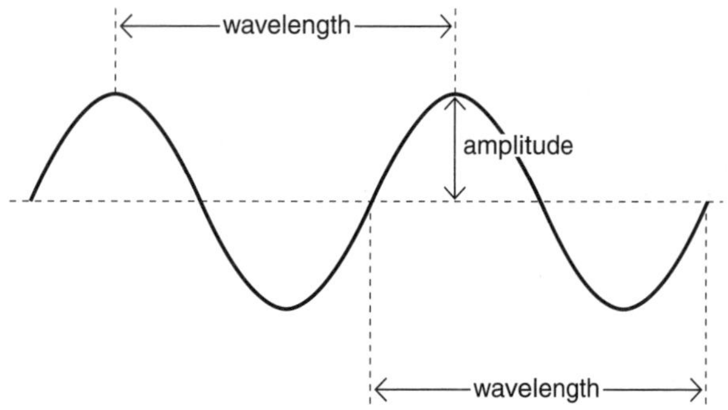

There are a number of features of a wave used for describing wave motion. The diagram above shows a transverse wave.

Displacement, x is the distance a point on the wave has moved from its undisturbed (rest) position.

It is measured in metres (m).

Amplitude, A or x_o is the maximum displacement of a point on the wave from its rest position

It is measured in metres (m).

Wavelength, λ is the distance from a point on the wave to the next point where the displacement is the same and the movement is in the same direction.

In the diagram above this is the distance between, for example, one crest and the next. It can be defined as the shortest distance between two points that are in phase (see right). It is measured in metres (m).

Period, T is the time for one complete wave to pass.

It is measured in seconds (s).

Phase difference, φ is a measure of how closely two waves with the same frequency are in step with each other.

If two points are 'in phase' or have 'zero phase difference' then the crests occur together. If two waves are exactly 'out of phase' or in 'antiphase' there is a difference of half a cycle, (or half a wavelength) between them.

The wave on the left is a sine wave, (it has the shape of a graph of $y = \sin x$), and one complete cycle on the graph represents a change in the angle x from 0 to 360° so phase difference can be expressed as an angle in degrees.

A phase difference $\varphi = 0°$ or $\varphi = 360°$ means waves are in phase.

A phase difference $\varphi = 180°$ means waves are out of phase.

Phase difference is also measured in radians (rad) where 2π radians in on cycle (2π rad = 360°) so a phase difference is always $0 \le \varphi < 2\pi$ rad. (If 2 waves are in phase, the phase difference is zero, because 2π rad is a complete cycle so the difference has reached zero again.) If two waves are in antiphase, the phase difference is π rad.

Frequency, f is the number of oscillations per second.

It is measured in hertz (Hz). A frequency of 1Hz is an oscillation of 1 cycle per second (1 Hz = 1 s^{-1}).

The frequency is determined by the rate that the source of the waves is oscillating, it is the number of waves emitted per second.

$$f = \frac{1}{T}$$

Speed of a wave, v (c for light in a vacuum) is the distance travelled by a point on the wave in unit time.

It is measured in metres per second (m s^{-1}).

The wave equation

For a point on a wave: speed, $v = \dfrac{\text{distance}}{\text{time}}$

In one period, T, the point travels a distance of one wavelength, λ

$$v = \frac{\lambda}{T}$$

period $T = \dfrac{1}{f}$

where f = the frequency of the wave. This gives:

$v = f\lambda$

which is called the **wave equation**.

For example, if the frequency is 5 Hz and the wavelength is 2 m, then the wave speed is $v = 5$ Hz $\times$ 2 m = 10 m s^{-1}.

Reflection, refraction and diffraction

Wave effects can be investigated using a **ripple tank** in which ripples travel across the surface of shallow water.

Reflection Waves striking an obstacle are reflected. The angle of incidence is equal to the angle of reflection.

The angle of incidence and the angle of reflection are measured to the normal, which is a line perpendicular to the surface (dotted in the diagram). The law of reflection holds for curved and rough surfaces, but in these cases the normals will be in different directions at different points on the surface.

All waves obey the law of reflection, including sound and light. Echoes are reflections of sound waves from hard surfaces. Mirrors reflect light and can form images. Rough surfaces reflect light in all directions.

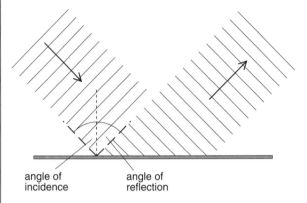

angle of incidence   angle of reflection

Refraction Refraction occurs when the waves slow down and is observed by the bending of the waves when the angle of incidence at the boundary is not zero. In a ripple tank, the waves can be slowed by using a flat piece of plastic to make the water shallower.

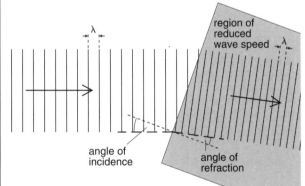

angle of incidence   region of reduced wave speed   angle of refraction

The frequency does not change when the wave changes speed, so it follows from the wave equation that the wavelength must change, it decreases as the wave speed decreases.

Light waves travel more slowly in a medium than in a vacuum. If they enter the medium with an angle of incidence greater than zero then one side of the wavefront is slowed down before the other side of the wave front. This results in the wave changing direction. The wave bends towards the normal (the angle of refraction is less than the angle of incidence) when light enters an optically denser medium (for example, from air to glass or air to water.) It speeds up and bends away from the normal when it enters a less dense medium (for example, from glass to air, or water to air).

Sound waves travel at different speeds in a different medium, for example in different gases, and can be refracted at the boundary between them, or between hot and cold air.

Diffraction Waves bend round the edges of a narrow gap. This is called *diffraction*. It is significant if the gap size is about a wavelength. Wider gaps cause less diffraction.

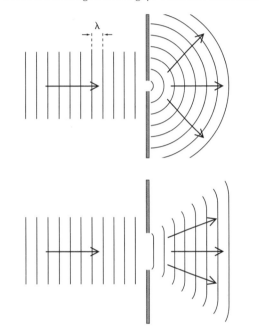

Notice that the wavelength does not change (the spacing of the lines stays the same).

The wavelength of sound in air is about a metre, so sound is diffracted through doorways.

The wavelength of light is about 5×10^{-7} m so the gap must be similar to a hairs width to see diffraction effects. Diffraction occurs with waves, but not with particles, so the diffraction effects that can be seen with light are good evidence that light can behave as a wave.

2.4.2 Electromagnetic waves

The electromagnetic spectrum

Light is one member of a whole family of transverse waves called the ***electromagnetic spectrum***.

All electromagnetic waves:

- Can travel through a vacuum.
- Are transverse waves.
- Are oscillations of a magnetic field and an electric field at right angles to each other.
- Travel through a vacuum at a speed of 3×10^8 m s^{-1}

Electromagnetic waves are emitted whenever electrons or other charged particles oscillate or lose energy. The greater the energy change, the lower the wavelength.

Wavelength in metres	Wave type	Typical sources, detection, uses, and effects
10^5		
	radio waves	From electrons oscillating in aerial. Detected by resonance in electronic circuits. Used for communication including TV.
3×10^{-2}		
	microwaves	From electrons oscillating in a magnetron, klystron oscillators. Detected by their heating effect, or electronic circuits. Used for radar, mobile phones, satellite navigation and microwave cooking.
10^{-3}		
	infrared	From all objects above the temperature of absolute zero. Detected by a thermopile, heating the skin, photographic film, night-shot digital cameras. Heating, night vision cameras and goggles, remote controls.
7.4×10^{-7}		
	visible light	From high temperature materials, lasers. Detected by the retina of the eye, photographic film, digital cameras. Used for sight, communication (especially using fibre optic cables).
4×10^{-7}		
	ultraviolet	From very hot objects. Detected by photographic film, phosphors, sunburn. Used for detecting forgeries, 'whiter than white' washing powders, disco lights, sun tanning.
10^{-8}		
	X-rays	From electrons stopped rapidly in X-ray tube. Detected by photographic film, fluorescence. Used in medicine (e.g. CT scans, X-ray photography,) crystal structure analysis. Causes ionisation so is hazardous to living things.
10^{-12}		
	gamma rays	From radioactive materials (differs from X-rays only in method of production) Detected by photographic film, Geiger Muler tube. Used in medicine (gamma camera, PET scans look for gamma rays emitted).
10^{-16}		

Ultraviolet radiation from the Sun

Ultraviolet radiation ranges from wavelengths of 10 nm to 400 nm. Most of the electromagnetic radiation from the Sun arriving at the top of the Earth's atmosphere has wavelengths longer than 100 nm, so the ultraviolet radiation arriving at the top of the Earth's atmosphere from the Sun is divided into 3 bands:

- UV-A Wavelengths 315 nm – 400 nm
- UV-B Wavelengths 280 nm – 315 nm
- UV-C Wavelengths 280 nm – 100 nm

UV-A accounts for approximately 95% of the UV radiation reaching the Earth's surface. It penetrates deeper into the layers of the skin and results in an immediate tanning effect. Furthermore, it also contributes to skin ageing and wrinkling. For a long time it was thought that UV-A could not cause any lasting damage, but recent studies strongly suggest that it may also contribute to the development of skin cancers.

The remaining (approximate) 5% of the UV radiation reaching the Earth's surface is UV-B. This is very biologically active but cannot penetrate beyond the surface skin layers. It is responsible for delayed tanning and burning. It also enhances skin ageing and there is conclusive evidence that it can cause skin cancer, especially melanoma. UV-B is normally absorbed by the ozone layer.

UV-C is the most damaging UV radiation from the Sun to living things. However, it is totally absorbed by the atmosphere and does not reach the Earth's surface.

Sunscreens

Most sunscreens contain a mixture of inorganic chemicals, such as titanium dioxide or zinc oxide, which reflect UV rays, and organic chemicals, which contain carbon molecules that absorb the UV rays. They prevent the UV rays reaching the skin.

In the UK sunscreens have a star rating, from 0 to 5 stars, to show how much UV-A they absorb. The Sun Protection Factor (SPF) indicates how much protection the product gives against UV-B.

Polarisation

In the diagram of progressive wave motion (see 'Progressive wave motion', page 57), the particles oscillate in a vertical *plane of vibration*. Light is usually a mixture of waves with different planes of vibration. It is **unpolarised**. Polarised filters transmits light in one plane of vibration only. Light like this is plane **polarised**.

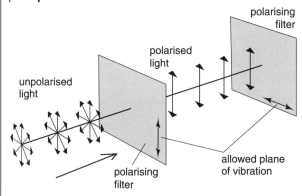

Above, polarised light from one polarising filter strikes a second. The light is blocked because its plane of vibration has no component in the allowed direction.

Malus' law

If polarised light with intensity I_0 is directed at a polarising filter at an angle θ to the allowed plane of vibration, the intensity θ of the light transmitted is

$$I = I_0 \cos^2 \theta$$

Only tranverse waves can be polarised. Experiments with polarising filters provide evidence that light waves are transverse.

Microwaves and radio waves generated in the laboratory are polarised, so the transmitter and receiver must be aligned.

Polarisation by reflection When an unpolarised light ray strikes the surface of a transparent medium such as water, the refracted ray is partly polarised. At most angles, the reflected ray is also partly polarised.

But if the reflected ray is at 90° to the refracted ray, it is *totally* polarised.

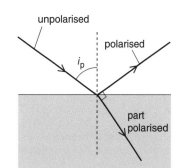

Polarising filter sunglasses reduce the glare from shiny reflective surfaces by blocking the reflected, polarised light.

2.4.3 Interference

Superposition and interference

Two sets of waves can pass through the same point without affecting each other. However, they have a combined effect, found by adding their displacements (as vectors). This is known as the **principle of superposition of waves**.

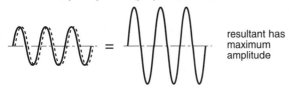

resultant has maximum amplitude

The waves above are *in phase* and reinforce each other. This is called **constructive interference**.

resultant has zero amplitude

The waves above have a **phase difference** of $\frac{1}{2}$ cycle (180° or π radians) and cancel each other. This is called **destructive interference**.

For **interference** to be observed:
- The sets of waves must be **coherent**: there must be a constant phase difference between them. For this, they must have the same frequency.
- The sets of waves must have approximately the same amplitude and plane of vibration.

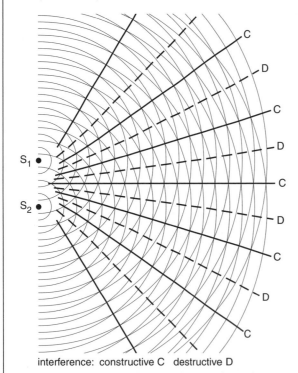

interference: constructive C   destructive D

Above, waves from two coherent sources, S_1 and S_2, produce regions of reinforcing and cancelling called an **interference pattern**. At each point of constructive interference, the path from one source is an exact number of wavelengths longer than from the other source (or the same length). The **path difference** is 0 or λ or 2λ, and so on. The **phase difference** is 0, 2π radians, 4π radians and so on.

An interference pattern can be produced using light waves. However, waves from separate sources are not normally coherent, so the two sets of waves must originate from the same source. Light of one frequency (and therefore of one wavelength and colour) is called **monochromatic light**. A laser emits monochromatic light which is coherent across its beam.

Intensity

Waves transmit energy. The energy transmitted per second by a beam of radiation is the power, P, measured in watts (W).

When the beam arrives at a surface, the intensity of the radiation depends on the power, the size of the surface and the angle that the beam is to the surface – the intensity will be greater if the surface is at right angles to the beam.

To calculate the **intensity, I** (in W m^{-2}):
$$I = \frac{P}{A}$$

On the right, waves are radiating uniformly from a source of power output P. At a distance r from the source, the power is spread over an area $4\pi r^2$. So intensity $I = P/4\pi r^2$.

Note that $I \propto 1/r^2$. This is an example of an inverse square law.

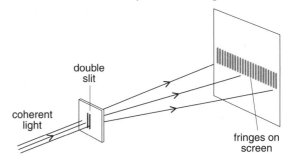

The Young double-slit experiment

A two-source interference experiment with light:

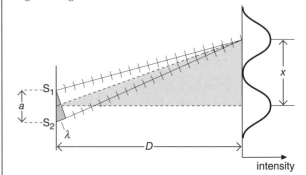

coherent light

double slit

fringes on screen

Above, light waves from a laser spread out by diffraction from two slits (typically less than $\frac{1}{2}$mm apart). The interference pattern produces a series of bright and dark **fringes** on the screen. The bright fringes are regions of constructive interference. The dark fringes are regions of destructive interference

The first bright fringe occurs where the path difference is λ. For small angles, the shaded triangles above are similar, and the following equation applies:
$$\frac{\lambda}{a} = \frac{x}{D}$$

So $\lambda = \dfrac{ax}{D}$ x is the **fringe spacing**

Note:
- The fringe spacing is increased if the slits are closer together or light of longer wavelength is used.
- By measuring x, D, and a, the wavelength of light can be found using the above equation. Light wavelengths range from 7×10^{-7} m (red) down to 4×10^{-7} m (violet).

The Young double-slit experiment *continued*

If one of the slits is covered up the pattern of fringes disappears because there is no longer interference between the waves from two coherent sources. Interference is a wave phenomenon and shows that light can behave as a wave.

The experiment can be done with a monochromatic light source such as a sodium lamp if a single slit is placed in front of the lamp and light from the single slit spreads out to go through the double slits. This ensures that the sodium light is coherent. Laser light is coherent, so the experiment can easily be set up by shining a laser at the double slits. This will result in a bright pattern of spots.

Two-source interference with microwaves:

Use a microwave transmitter that produces waves with wavelength of about 3 cm. Set up the experiment in a similar way to the light experiment in the diagram above. Use three sheets of aluminium as barriers and arrange them so that the 'slits' are about half a wavelength wide.

Move the microwave receiver in a semicircle so that it is always the same distance from the transmitter. The receiver should detect about 5 maxima.

When either slit is covered with another metal plate the interference pattern disappears, so at points where there was constructive interference and a maximum signal, the signal is now less, and where there was destructive interference and the signal was a minimum, the signal is now greater.

Two-source interference of sound waves:

Two speakers are connected to the same signal so the sound is coherent.

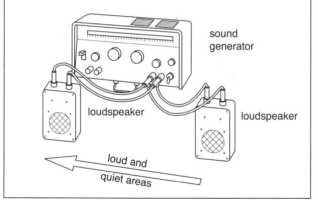

Amplitude

A wave with larger amplitude transmits more energy than one with a lower amplitude. The amplitude is the maximum displacement of the particles in the wave from their rest position, and this depends on the energy that is given to the particles by the source of the waves.

Energy in a wave is proportional to (amplitude)^2.

From the equation for intensity page 62, and using the fact that power is the energy arriving per unit time, gives:

Intensity is proportional to (amplitude)^2.

Diffraction grating

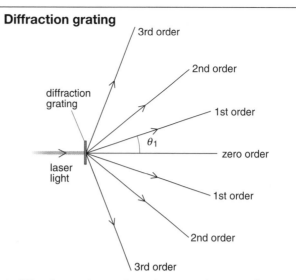

A **diffraction grating,** as above, has many slits (typically, 500 per mm). Constructive interference produces sharp lines of maximum intensity at set angles either side of a sharp, central maximum. In between, destructive interference gives zero or near-zero intensity. To identify the lines, they are each given an **order number** (0, 1, 2 etc.).

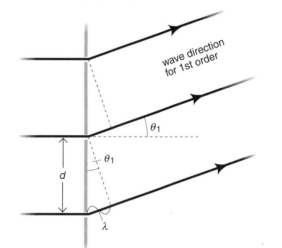

A close-up of part of the grating is shown above. d is the **grating spacing**. θ_1 is the angle of the first order maximum. In this case, the path difference for any two adjacent slits is one wavelength, λ. From the triangles,

$$\sin \theta_1 = \frac{\lambda}{d} \qquad \text{So } d \sin \theta_1 = \lambda$$

For higher orders, the path differences are 2λ, 3λ etc. The following equation gives values of θ for all orders:

$$d \sin \theta = n\lambda$$

where n is the order number (0, 1, 2 etc.).

Note:
- If d, θ_1, and n are known, λ can be calculated.
- A longer wavelength gives a larger angle for each order.
- If the incoming light is a mixture of wavelengths (e.g. white), each order above zero becomes a spectrum.

Using a diffraction grating with many slits, rather than two slits, means that

- the intensity of the maxima will be much greater, because the effect of all the slits adds up giving brighter maxima.
- the maxima are more sharply defined.
- using a diffraction grating in which the 'slits' are closer together means that the angles are larger so they can be measured with greater precision.

2.4.4 Stationary waves

Formation of stationary waves

Stationary waves are produced by the superposition of two sets of progressive waves (of equal amplitude and frequency) travelling in opposite directions. For example, when a stretched string is vibrated, waves travel along the string, reflect from the ends and are superimposed on waves travelling the other way.

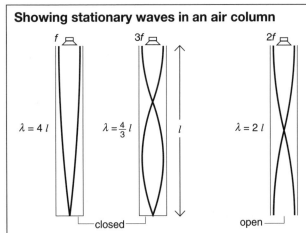

wave travelling to the right
wave travelling to the left
resultant stationary wave

At $t = 0$ and $t = \frac{1}{2} T$ the resultant is zero at all points

At $t = \frac{1}{4} T$ and $t = \frac{3}{4} T$ the displacement at all points is the maximum value, which varies from 0 to 2 A where A is the amplitude of the two progressive waves.

Nodes are the points where the displacement is always zero.

Antinodes are the points where the displacement varies from maximum positive value through zero to maximum negative value and back.

The distance between two adjacent nodes $= \frac{\lambda}{2}$

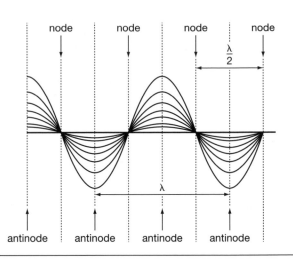

Showing stationary waves in an air column

If a sound source is placed near the open end of a pipe, there are certain frequencies at which stationary waves are set up in the air column and the sound intensity reaches a maximum. Three examples are shown above.

Note:
- Sound waves are *longitudinal*. The 'waves' in each pipe are a *graphical representation* of the amplitude, where amplitude is the resultant maximum displacement of the air in the pipe.
- Where the end of a pipe is open, there is an antinode. Where the end of a pipe is closed, there is a node.

Stationary and progressive waves

This table summarises the similarities and differences between progressive waves and stationary waves

	Stationary wave	Progressive wave
velocity	Does not travel.	Travels with wave velocity v.
Energy	Energy is stored in the wave but is not transmitted.	Energy is carried in the direction of travel of the wave.
Frequency	All particles vibrate with frequency f except those at the nodes which are at rest.	All particles vibrate with frequency f.
Wavelength	2 × distance between a pair of adjacent nodes or antinodes	Distance between adjacent particles which have the same phase (e.g. 2 crests).
Amplitude	Varies from zero at the nodes (which are always at rest) to a maximum of 2 A at the antinodes.	Constant value A for all particles in the wave.
phase	All particles between two adjacent nodes have the same phase. All particles on one side of a node have a phase difference of π rad from those on the other side of the node.	All the particles within one wavelength have different phase.

Showing stationary waves with microwaves

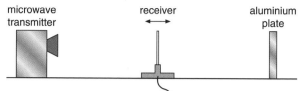

microwave transmitter   receiver   aluminium plate

to meter or amplifier and loudspeaker

The microwaves are emitted from the transmitter and reflected back from the aluminium plate to produce a standing waves pattern between the two. To set up the standing waves the position of the transmitter is adjusted so that there is a signal at the receiver, which is positioned between the plate and the transmitter.

Moving the receiver along the line between the aluminium plate and the transmitter the amplitude of the signal will rise and fall. It will be zero at the nodes and a maximum value at the antinodes.

To find the wavelength of the microwaves measure the distance between the first and last antinode and use this to calculate the distance between two adjacent antinodes. This will be half the wavelength.

Showing stationary waves in a stretched string

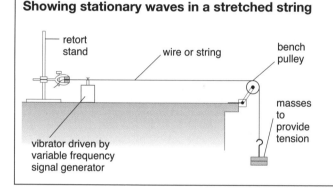

retort stand   wire or string   bench pulley

masses to provide tension

vibrator driven by variable frequency signal generator

The vibrator is switched on and the amplitude increased. As the frequency of the vibrator is adjusted stationary wave patterns will appear at certain frequencies, some of these **modes** are shown below. A stroboscope can be used to freeze the motion of the string and make it easier to see the patterns.

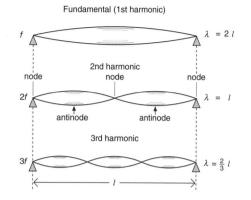

Fundamental (1st harmonic)

f $\lambda = 2l$

2nd harmonic

node   node   node

$2f$ $\lambda = l$

antinode   antinode

3rd harmonic

$3f$ $\lambda = \frac{2}{3}l$

l

The lowest frequency stationary wave is called the **fundamental mode** or the **1st harmonic**. The length of the string is half the wavelength of the sound.

The next frequency that produces standing waves is twice the fundamental frequency and is called the **2nd harmonic**. The length of the string is equal to the wavelength and there is a node in the centre of the string.

Further harmonics follow the same pattern, as shown in the diagram.

Example

A stretched string 80 cm long is set in vibration. What is the wavelength of the fundamental mode?

There is a node at each one of the string, so $l = \lambda/2$

$0.80\ \text{m} = \lambda \div 2$

$\lambda = 1.60\ \text{m}$

The speed of sound

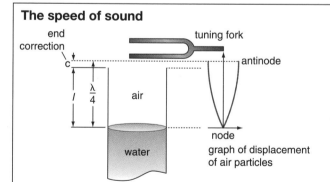

end correction

c

l $\frac{\lambda}{4}$ air

tuning fork

antinode

node

water   graph of displacement of air particles

A tuning fork is held over the end of a tube filled with water and the level of the water is gradually lowered. A stationary wave is set up when the level is low enough for an air column one quarter of the wavelength to be set in motion. The air slightly outside the pipe is also set in motion, so that the antinode is slightly above the top of the pipe, this is called the end correction, c. If the length from the top of the tube to the water level is l, and the

$\frac{\lambda}{4} = l + c$

The velocity of sound in air is v, and the frequency of the tuning fork is f, so $\lambda = v/f$

This gives an equation:

$l + c = \dfrac{v}{4f}$

This equation can be written:

$l = \dfrac{v}{4f} - c$

If tuning forks of known frequency, f are used and the length l, is measured, a graph of l plotted against $1/f$ will be a straight line with gradient $v/4$ and intercept on the y axis of $-c$.

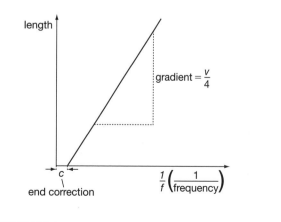

length

gradient $= \dfrac{v}{4}$

c

end correction

$\dfrac{1}{f}\left(\dfrac{1}{\text{frequency}}\right)$

Module 5: Quantum physics

2.5.1 Energy of a photon

Photons and quantum energy

Some effects suggest that light has particle-like properties. These include the **photoelectric effect**. Einstein (in 1905) was able to explain this by assuming that light (or other electromagnetic radiation) is made up of 'packets' of wave energy, called **photons**. He used the theory that Planck put forward in 1900 that energy cannot be divided into smaller and smaller amounts. It is only emitted in discrete 'packets', each called a **quantum**. The energy E of a quantum depends on the frequency f of the radiating source, as given by this equation:

$$E = hf$$

where h is known as the **Planck constant**. Its value, found by experiment, is 6.63×10^{-34} J s.

A **photon** is a quantum of energy of electromagnetic radiation

For electromagnetic radiation, $c = f\lambda$, so the equation can be rewritten as

$$E = \frac{hc}{\lambda}$$

where c is the speed of light and λ the wavelength.

Note:
- The shorter the wavelength (and therefore the higher the frequency), the greater the energy of each quantum.
- A quantum is an extremely small amount of energy.

Producing an electron beam

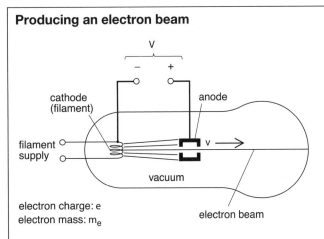

electron charge: e
electron mass: m_e

In the vacuum tube above, electrons are given off by a hot tungsten filament. The effect is called **thermionic emission**. The electrons gain kinetic energy (KE) as they are pulled from the **cathode** (–) to the **anode** (+).

Some pass through the hole in the anode and emerge as a narrow beam, at speed v. Electrons in a beam are sometimes called **cathode rays** because they come from the cathode.

As an electron (charge e) moves from cathode to anode,

KE gained = work done

W = VQ

So $\frac{1}{2} m_e v^2 = eV$

Electron gun This is a device that produces a narrow beam of electrons. It uses the principle described above. In many electron guns, the cathode is an oxide-coated plate, heated by a separate filament.

Electronvolt (eV) This is a unit often used for measuring particle energies. 1 eV is the energy gained by an electron when moving through a p.d of 1 V.

If an electron (1.60×10^{-19} C) moves through 1 V,
    KE gained = charge × p.d = $1.60 \times 10^{-19} \times 1$ J
So $1 \text{ eV} = 1.60 \times 10^{-19}$ J

Example

Electrons are accelerated from rest through a p.d. of 500 V. What is their speed?

$$eV = \frac{1}{2} m v^2$$

$$1.6 \times 10^{-19} \text{ C} \times 500 \text{ V} = \frac{1}{2} \times 9.11 \times 10^{-31} \text{ kg} \times v^2$$

$$v = \sqrt{(1.6 \times 10^{-16} \div 9.11 \times 10^{-31})} = 1.3 \times 10^7 \text{ m s}^{-1}$$

Note: As the particles approach the speed of light $c = 3 \times 10^8$ m s^{-1} there will be relativistic effects and the equation no longer holds, it will give an answer greater than the speed of light. The particles cannot travel faster than light and they will increase in mass.

An experiment to estimate the Planck constant

A light emitting diode (LED) emits a photon when a quantum of energy is transferred from an electron passing through it.

Energy lost by an electron = charge on electron × p.d. across LED

$E = eV$

When all this energy is transferred to the photon $E = hf = \dfrac{hc}{\lambda}$

Where h is the Planck constant, c is the speed of light, and λ is the wavelength of light emitted by the LED.

Using the circuit shown, the p.d. is measured across an LED when it just begins to glow. This will be when the energy of the electron is just enough to release a photon, so that all the energy of the electron is transferred to the photon

$\dfrac{hc}{\lambda} = eV$

A number of different colour LEDs are used. Their wavelength can be found from the manufacturers data sheets. The LEDs must be single colour LEDs and not, for example yellow ones that use a green and red LED to produce the yellow light.

A graph of V against $\dfrac{1}{\lambda}$ will be a straight line with gradient $\dfrac{hc}{e}$.

gradient, $m = \dfrac{(2.14-1.10)}{(2.40-1.60) \times 10^6} = 1.30 \times 10^{-6}$

$h = \dfrac{me}{c} = \dfrac{1.30 \times 10^{-6}\ \text{Vm}^{-1} \times 1.60 \times 10^{-19}\text{C}}{3.00 \times 10^8\ \text{ms}^{-1}} = 6.93 \times 10^{-34}\ \text{J s}$

(the actual value of the Planck constant is $h = 6.63 \times 10^{-34}\ \text{J s}$).

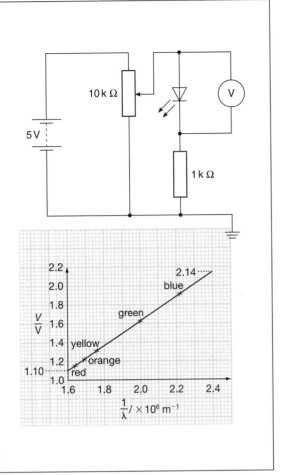

2.5.2 The photoelectric effect

The photoelectric effect

When some substances are illuminated by light (or shorter wavelengths), electrons are emitted from their surface. This is called the **photoelectric effect**. The electrons are emitted instantaneously with a range of kinetic energies, up to a maximum.

Experiments show that:
- Increasing the intensity of the light increases the number of electrons emitted per second.
- For light beneath a certain **threshold frequency**, f_0, no electrons are emitted, even in very intense light.
- Above f_0, the maximum E_k of the electrons increases with frequency, but is not affected by intensity. Even very dim light gives some electrons with high E_k.

The wave theory cannot explain the threshold frequency, or how low-amplitude waves can cause high-E_k electrons. The photoelectric effect provides good evidence that electromagnetic waves can behave as particles.

Einstein's quantum explanation Energy is conserved when a photon interacts with an electron. Each photon delivers a quantum of energy, hf, which is absorbed by an electron. Energy Φ is needed to free the electron from the surface. If hf is more than this, the remainder is available to the electron as E_k (though most electrons lose some E_k before emission because they interact with other atoms). So

$$hf \quad = \quad \Phi \quad + \quad \tfrac{1}{2}m_e v_{max}^2 \qquad (1)$$

| energy delivered by photon | minimum energy needed to free electron from surface | E_k of electron (with no further energy losses) |

The maximum kinetic energy of the electrons is independent of the intensity of the light is because it depends on the energy of the photon the electron has absorbed and increasing the number of photons will not affect the energy of one photon.

Investigating the photoelectric effect

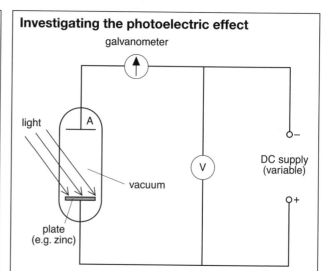

The principle of an experiment to investigate the photoelectric effect is shown above. The material being investigated (e.g. zinc) is illuminated with light of known frequency, f. Emitted electrons reach plate A, so the galvanometer detects a current in the circuit. The maximum E_k of the emitted electrons is found by applying just enough *opposing* voltage, V_s, to *stop* them reaching A, so that the galvanometer reading falls to zero.

V_s is called the **stopping voltage**. At this voltage

$$eV_s = \tfrac{1}{2}m_e v_{max}^2$$

So, if equation (2) below is correct,

$$eV_s = hf - hf_0$$

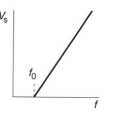

Therefore, if V_s is measured for light of different frequencies, a graph of V_s against f should be of the form shown above.

The photoelectric current in a photocell circuit is proportional to the intensity of the incident radiation. This is because, once the threshold frequency is exceeded, increasing the intensity increases the number of incident photons. This increases the number of electrons which each absorb one photon and escape from the metal plate in the photocell. The flow of electrons forms the current.

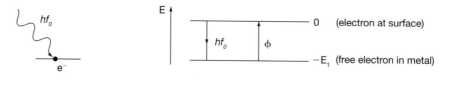

photon of frequency f_0 has just enough energy to free an electron from the metal

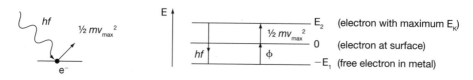

photon of frequency f causes emission of electrons with maximum E_k $E_2 = \tfrac{1}{2}mv_{max}^2$

Note:
- The number of electrons emitted is proportional to the number of photons absorbed.
- Φ is called the **work function**. Materials with a low Φ emit electrons in visible light. Those with a higher Φ require the higher-energy photons of ultraviolet.

- If $hf < \Phi$, no electrons are emitted.
- The energy of a photon at the threshold frequency $= hf_0$ $= \Phi$. So, equation (1) can be rearranged and rewritten:

$$\tfrac{1}{2}m_e v_{max}^2 = hf - hf_0 \qquad (2)$$

2.5.3 Wave–particle duality

Wave–particle duality

Light waves have particle-like properties. De Broglie (in 1922) suggested that the converse might also be true: matter particles, such as electrons, might have wave-like properties. There might be **wave–particle duality**.

According to de Broglie, if a particle of momentum mv is associated with a **matter wave** of wavelength λ, then

$$\lambda = \frac{h}{mv}$$

If a beam of electrons is passed through a thin layer of graphite, the electrons form a diffraction pattern. This suggests that the rows of atoms are acting rather like a diffraction grating and the electrons are behaving as waves.

Electron diffraction

In an electron diffraction tube electrons are accelerated and hit a thin film of polycrystalline graphite. The graphite behaves in a similar way to a diffraction grating. The waves are diffracted by atoms and the spaces between the atoms in the graphite and produce an interference pattern. The pattern observed with an electron diffraction tube is two bright concentric rings. These are produced by two different spacings of atomic layers in the graphite structure.

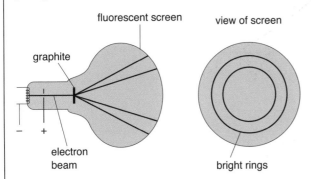

fluorescent screen   view of screen

graphite

electron beam   bright rings

At a point where the electron waves interfere constructively the amplitude is a maximum. This means that there is a high probability of an electron arriving at that point. Lots of electrons arrive there and the screen coated with a phosphor (such as zinc sulphide) glows brightly.

At a point where the amplitude is zero, there is no chance of an electron arriving there so no light is emitted.

Typical de Broglie wavelengths of electrons

An electron accelerated through 2000 V produces a first order ring at an angle of 8.62°.

The spacing of the atoms in the graphite can be calculated using the diffraction grating equation

$d \sin\theta = n\lambda$ where d is the atom spacing, $n = 1$ for the 1st order, $\theta = 8.94°$ and λ is the de Broglie wavelength of the electrons.

$$\lambda = \frac{h}{mv} \quad \text{and} \quad eV = \frac{1}{2} mv^2$$

so $v = \sqrt{(2\,eV \div m)} = \sqrt{(2 \times 1.6 \times 10^{-19}\ C \times 2000\ V \div 9.11 \times 10^{-31}\ kg)} = 2.6 \times 10^7\ m\,s^{-1}$

and $\lambda = \dfrac{6.63 \times 10^{-34}\ J\,s}{(9.11 \times 10^{-31}\ kg \times 2.6 \times 10^7\ m\,s^{-1})} = 2.80 \times 10^{-11}\ m$

This gives $d \sin(8.94°) = 1 \times 2.80 \times 10^{-11}$ m.

The atomic spacing $d = 1.80 \times 10^{-10}$ m.

The short wavelength is the reason why electron diffraction cannot be observed using ordinary diffraction gratings. The spacing of the 'slits' needs to be very small. Because atoms have diameters of the order of 10^{-10} m the layers of atoms in the graphite behave like a diffraction grating with a small enough spacing.

As well as finding the atomic spacing, the diffraction pattern can be used to find the arrangement of atoms in crystals. As the accelerating voltage is increased the wavelength gets smaller and high energy electrons can be used to investigate structures with smaller spacings; they can be used to determine the size of nuclei.

2.5.4 Energy levels in atoms

Spectral lines

A **spectrum** contains a spread of wavelengths, but not always in a continuous range. For example, if there is an electric discharge through hydrogen at low pressure (such as in a gas discharge lamp), the gas emits particular wavelengths only, so the spectrum is made up of lines (visible colours, ultraviolet and infrared), some of which are shown below:

Hydrogen

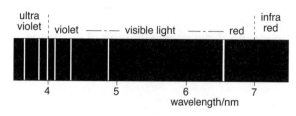

When electrons are bound in an atom they can only have certain energies. These are represented as **energy levels**.

When an electron gains energy, for example by absorbing a photon of radiation or colliding with another particle, it may be raised to a higher energy level. It can only absorb the exact amount of energy required to move to another allowed level, or enough energy to leave the atom.

When it drops back to a lower energy level the electron loses the exact amount of energy corresponding to the energy difference between the levels, this is called a quantum of energy. For example in dropping from energy level E_1 to energy level E_2 it loses energy $(E_1 - E_2)$, which is emitted as a photon. So

$$hf = E_1 - E_2$$

and $\dfrac{hc}{\lambda} = E_1 - E_2$

Emission and absorption spectra

If light is radiated directly from its source, its spectrum is called an **emission spectrum**. Examples include the line spectrum above and the continuous spectrum of the Sun.

The Sun's emission spectrum is crossed by many faint, dark lines. These are an **absorption spectrum**. They occur because some wavelengths emitted by the Sun's core are absorbed by cooler gases (e.g. hydrogen) in its outer layers.

Some of the lines in the absorption spectrum of hydrogen are shown below. When the Sun's radiation passes through the gas, the atoms *absorb* photons whose energies match those in their emission spectrum. They then re-emit photons of these energies, but in all directions, so the intensity in the forward direction is reduced for those wavelengths.

Hydrogen (absorption)

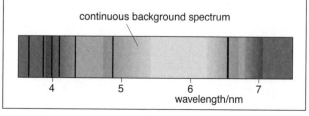

An electron jump is called a **transition**.
- The greater the energy change $(E_1 - E_2)$ of the transition, the higher the frequency f of the photon.
- Each possible transition gives a different spectral line.

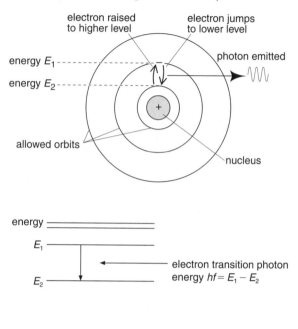

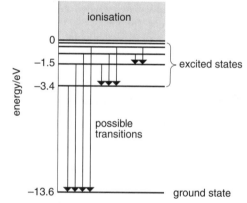

Example

Energy levels for the hydrogen atom:

A line spectrum is a feature of any gas in which individual atoms do not interact. If atoms exert forces on each other, many more energy levels are created. Tightly-packed atoms or molecules which are vibrating, rotating, or colliding with each other have so many possible energy states that the spectrum is a continuous range of colours.

The main energy levels and transitions for hydrogen (with isolated atoms) are shown above.

Note:
- If an atom is in its ground state, no electron has an unoccupied energy level beneath it.
- If an atom is in an excited state, an electron has been raised to a higher energy level, so there is an unoccupied level beneath it.
- If an atom is in an ionised state, an electron has been raised above the highest energy level (i.e. it has escaped). From the energy scale on the above chart, the minimum energy required to ionise a hydrogen atom is 13.6 eV.

Self-assessment questions

After revising a section you should try these questions.

Questions are only given for those sections which relate to compulsory material in the specifications.

Answers, including references to sections where you can find more detail, begin on page 150.

You should review the work relating to any questions that you were unable to do or when you obtain incorrect numerical answers.

Where necessary assume that the acceleration of free fall $g = 9.81$ m s^{-2}.

Unit 1 Module 1

1. Write down the following quantities in standard form:
   (a) 3.5 MΩ **(b)** 15 mm s^{-1} **(c)** 25 mm^2
2. Convert the following quantities to include a number with a suitable prefix:
   (a) 5.0×10^7m **(b)** 3.2×10^{-3} A **(c)** 39×10^{-9} s
3. **(a)** Write down the defining equation for pressure.
   (b) Use the defining equation to arrive at a unit for pressure in terms of base units.
4. A quantity is quoted as $(3.7 \pm 0.2) \times 10^{-3}$ m. Calculate the percentage uncertainty in this quantity.
5. The diameter of a wire is measured as 1.2 ± 0.1 mm. The length of the wire is 73.0 ± 0.5 cm.
   Determine
   (a) the volume of the wire
   (b) the uncertainty in the volume.
6. State the difference between a vector quantity and a scalar quantity.
7. A student runs round a circular track of radius 40 m in 30 s. Calculate the average speed of the student.
8. The graph shows how the speed of a car varies with time from the instant when a driver sees a dog running into the road:

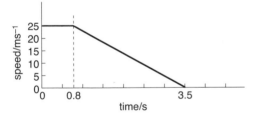

   (a) Determine the distance travelled before the driver applies the brakes.
   (b) Calculate the deceleration produced when the brakes are applied.
9. The graph shows how the speed of an athlete varies with time during the run-up to a long jump.

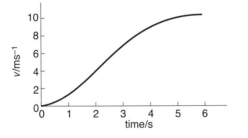

   Estimate
   (a) the maximum acceleration of the athlete
   (b) the length of the run-up.
10. Assuming no air resistance, how long would it take a ball to fall a distance of 100 m from rest?

11. A car travelling at 10 m s^{-1} accelerates uniformly at 2.5 s^{-2} for 3.5 s.
    (a) How far does it travel while accelerating?
    (b) Calculate its final speed.
12. A dart player throws a dart that hits the dartboard at the same height as that from which it was thrown.
    The dart was thrown from a distance of 3.0 m from the board and takes 0.25 s to reach the board.
    Calculate
    (a) the maximum height reached by the dart
    (b) the horizontal speed of the dart
    (c) the vertical speed of the dart when it reaches the dartboard.

Unit 1 Module 2

1. **(a)** State the difference between *mass* and *weight*.
   (b) State the unit in which each is measured.
2. Determine the resultant force in each of the following:

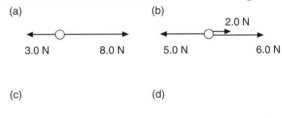

3. State the two conditions that are necessary for a system to be in equilibrium.
4. A force of 250 N is applied to a box at an angle of 35° to the horizontal. Calculate the horizontal and vertical components of this force.
5. The three strings/shown in the diagram are in tension, and in equilibrium.
   Calculate the tension in each of the strings.

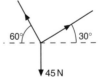

6. The diagram shows one arm of a 'mobile'. Calculate the weight of the ship.

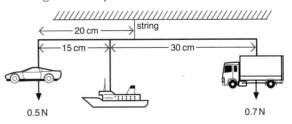

7. State the principle of moments.

8. A person of weight 550 N stands on a uniform plank of weight 50 N in the position shown in the diagram:

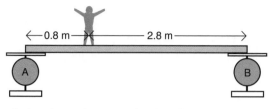

Calculate the reading in N of each of the scales A and B.

Unit 1 Module 3

1. Calculate
   (a) the kinetic energy of a car of mass 850 kg travelling at 110 km h^{-1}
   (b) the change in potential energy when a skier of mass 70 kg skis from a height of 1500 m above sea level to a height of 1210 m above sea level.

2. A weight lifter lifts a mass of 110 kg through a height of 0.500 m in 0.400 s. Calculate the useful power developed during the lift.

3. A van that develops an output power of 3.0 kW moves at a steady velocity of 30 m s^{-1}.
   (a) Calculate the total forward force developed between the tyres and the road.
   (b) Explain why the total drag forces must be exactly equal to the forward force in this case.

4. The kinetic energy of a body changes from 120 J to 40 J. It travels 3.5 m as this change occurs. Calculate the force acting on the body.

5. A piano wire made of steel (Young modulus = 2.0×10^{11} Pa) has a length of 1.30 m and a diameter of 2.0 mm. It stretches 2.5 mm when it is tightened. Calculate the tension in the wire.

6. Give an example of
   (a) a brittle material
   (b) a ductile material

7. A tendon stretches 3.5 mm when the tension is 14 N. It obeys Hooke's law up to this tension. Calculate the energy stored in the tendon.

8. Draw a diagram showing how the strain varies with stress for rubber. Explain what is happening during the different stages at a molecular level.

9. What makes brittle materials brittle?

10. Marble is strong in compression and has a Young modulus of 5.0×10^{10} Pa. A pillar of length 5.0 m and cross-sectional area 1.5 m^2 supports a load of 20 000 kg. Calculate
   (a) the stress in the pillar
   (b) the strain in the pillar
   (c) the compression of the pillar.

Unit 2 Module 1

1. Calculate the number of electrons passing each point in a circuit in 30 s when the current is 30 mA.

2. A copper wire of diameter 0.80 mm carries a current of 1.5 A. The copper contains 8×10^{28} electrons per m^3. Determine the drift speed of the electrons down the wire.

Unit 2 Module 2

1. A car headlamp is connected to a 12.0 V supply of negligible internal resistance. The lamp then works at its rated power of 24 W. Calculate
   (a) the current in the circuit
   (b) the resistance of the filament when the lamp is working normally.

2. When a potential difference of 9.0 V is placed across a wire there is a current of 1.6 A in the wire. The wire has a length of 1.8 m and an area of cross-section of 0.15 mm^2. Calculate the resistivity of the material from which the wire is made.

Unit 2 Module 3

1. **(a)** Explain what is meant by a supply having an e.m.f. of 12 V and an *internal resistance* of 0.5 Ω.
   (b) Calculate the current in a 4.0 Ω resistor when it is connected between the terminals of the supply.
   (c) Determine the current that would flow if the supply terminals were short circuited.

2. Calculate the total resistance of each of the following combinations of resistors:

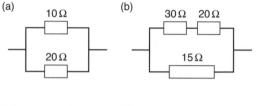

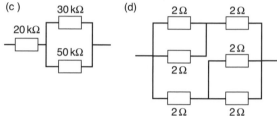

3. **(a)** Two components of resistance 4.0 Ω are connected in series to a 5.0 V supply of negligible internal resistance. Calculate
   (i) the current in the circuit
   (ii) the power dissipated by each resistor.
   (b) The components are now connected in parallel to the same supply. Calculate **(i)** the new current and **(ii)** power dissipated for each resistor.

4. **(a)** Determine the readings of the meters in the following circuit:

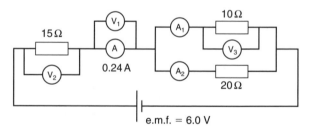

   (b) State the terminal potential difference of the supply.
   (c) Explain why this is lower than the e.m.f. of the supply.

5. **(a)** Determine the magnitude of the output voltage in the circuit below.

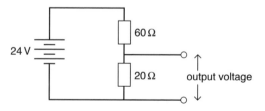

   (b) A 4.0 Ω resistor is connected across the output terminals. Determine the new value of the output voltage.
   (c) Determine the current drawn from the supply in **(a)** and **(b)**.

Unit 2 Module 4

1. Draw a sketch to show how the displacement of the medium varies with distance from the source for a sinusoidal oscillation. Indicate on your sketch the wavelength and the amplitude of the oscillation.

2. Distinguish between
   (a) a longitudinal and a transverse mechanical wave
   (b) a progressive and a stationary wave
   (c) a node and an antinode in a stationary wave.

3. Place the following parts of the electromagnetic spectrum in order of increasing frequency.
   ultraviolet, UHF radio, gamma rays, visible light, microwaves

4. In an experiment, it is found that it is possible to polarise the radiation used. State and explain the conclusions that can be drawn from this observation.

5. The intensity of the radiation reaching the outer atmosphere of the Earth is 1400 W m^{-2}. Calculate the intensity on Pluto.
   Distance from Earth to Sun = 1.5×10^{11} m.
   Distance from Pluto to Sun = 5.9×10^{12} m.

6. (a) Explain what is meant by superposition of waves.
   (b) State the conditions necessary for observation of an interference pattern from two sources.

7. Explain the role of diffraction in setting up an experiment to observe interference using light.

8. Calculate the fringe spacing produced as a laser light of wavelength 650 nm, passes through two slits separated by 1.5 mm, at a distance of 2.5 m.

9. An observer notices that if she walks 1.8 m in the direction shown in the diagram, the sound intensity falls from a maximum to a minimum. The speakers are connected to the same output of a frequency generator and are emitting notes of the same frequency.

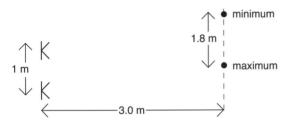

   (a) Explain this observation.
   (b) By means of a scale diagram determine the wavelength of the sound.

10. A diffraction grating has 300 slits per mm. Calculate the angular separation of the second-order maxima for red light (wavelength = 650 nm) and blue light (wavelength = 450 nm).

11. A string resonates at a fundamental frequency of 200 Hz. The length of the string is 0.65 m.
    (a) Sketch a diagram to show this mode of vibration.
    (b) Calculate the frequency and wavelength of the third harmonic vibration for this string.

12. A pipe that is closed at one end has a length of 0.45 m. It resonates at a fundamental frequency of 190 Hz. Estimate the speed of sound in air.

Unit 2 Module 5

1. The work function of zinc is 4.31 eV.
   (i) Calculate the energy in J of an electron emitted from a zinc target by UV radiation of wavelength 200 nm.
   (ii) Calculate the threshold frequency for zinc.

2. Calculate the frequency of the radiation emitted when an electron in a hydrogen atom undergoes a transition from the level at –1.5 eV to the ground state at –13.6 eV.

3. A carbon dioxide laser emits infrared radiation of wavelength 10.6 μm. Calculate the difference between the energy levels that give rise to this emission.

4. Calculate the wavelength of a photon of energy 2.2 eV.

5. Calculate the photon energy for radiation of wavelength 1.50×10^{-10} m.

Self-assessment answers

Unit 1 Module 1

1. (a) $3.5 \times 10^6\,\Omega$ (b) $1.5 \times 10^{-3}\,\text{m s}^{-1}$
   (c) $2.5 \times 10^{-5}\,\text{m}^2$
2. (a) 50 Mm   (b) 3.2 mA   (c) 39 ns
3. (a) $P = F/A$ (b) $\text{kg m}^{-1}\,\text{s}^{-2}$
4. 5.4%
5. (a) $8.3 \times 10^{-7}\,\text{m}^3$
   (b) 17.4% or absolute uncertainty = $\pm 1.4 \times 10^{-7}\,\text{m}^3$
   (so answer = $8\,(\pm 1.4) \times 10^{-7}\,\text{m}^3$)
6. Vector quantity has magnitude (size) and direction.
   Scalar quantity has magnitude only.
7. $8.4\,\text{m s}^{-1}$
8. (a) 20 m   (b) $9.3\,\text{m s}^{-2}$
9. (a) $\sim 3\,\text{m s}^{-2}$ (b) ~ 36 m
10. 4.52 s
11. (a) 50 m   (b) $19\,\text{m s}^{-1}$
12. (a) 0.31 m   (b) $12\,\text{m s}^{-1}$
    (c) $2.5\,\text{m s}^{-1}$

Unit 1 Module 2

1. Mass is the quantity of matter in a body (scalar).
   Weight is the force of attraction of the Earth on the mass (vector).
2. (a) 5 N to the right   (b) 3 N to the right
   (c) 9.4 N at 32° to horizontal
   (d) 20.5 N at 15° to horizontal (by scale drawing)
3. No resultant moment (torque) and no resultant force.
4. Horizontal force = 205 N; vertical force = 143 N
5. $T_1 = 39$ N, $T_2 = 22$ N, 45 N
6. 2.2 N
7. See 1.2.3 Equilibrium.
8. A reads 503 N; B reads 197 N.

Unit 1 Module 3

1. (a) $4.0 \times 10^5\,\text{J}$ (b) $2.0 \times 10^5\,\text{J}$
2. 1350 W
3. (a) 100 N
   (b) When velocity is constant there is no acceleration.
   This is the case only when there is no resultant force.
4. 23 N
5. 1200 N
6. (a) glass   (b) copper
7. 0.025 J
8. See 1.3.4 Behaviour of springs and materials.
9. See 1.3.4 Behaviour of springs and materials.
10. (a) $1.3 \times 10^5\,\text{Pa}$
    (b) 2.6×10^{-6}
    (c) $1.3 \times 10^{-5}\,\text{m}$

Unit 2 Module 1

1. 5.6×10^{18} electrons
2. $2.3 \times 10^{-4}\,\text{m s}^{-1}$

Unit 2 Module 2

1. (a) 2.0 A   (b) $6.0\,\Omega$
2. $4.7 \times 10^{-7}\,\Omega\,\text{m}$

Unit 2 Module 3

1. (a) An e.m.f. of 12 V means that 12 J of work is done when 1 coulomb of charge passes around the complete circuit. The resistance of the components inside the supply (chemicals, wires, etc.) is 0.5 Ω.
   (b) 2.7 A
   (c) 24 A
2. (a) $6.7\,\Omega$ (b) $12(2\text{sf})\,\Omega$ (c) $39\,\text{k}\Omega$
   (d) $1.5\,\Omega$

3. (a) (i) 0.63 A   (ii) 1.6 kW (2sf)
   (b) (i) 1.3 A (2sf)   (ii) 6.3 W
4. (a) $V_1 = 0$; $V_2 = 3.6$ V; $V_3 = 1.6$ V; $A_1 = 0.16$ A; $A_2 = 0.08$ A
   (b) 5.2 V
   (c) Some e.m.f. is 'lost' in producing current in the internal resistance of the supply.
5. (a) 6.0 V   (b) 1.26 V = 1.3 V (2sf)
   (c) 0.3 A in (a); 0.38 A in (b)

Unit 2 Module 4

1.
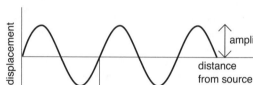

2. (a) Longitudinal: particles of medium oscillate in direction of energy transfer.
   Transverse: particles of medium oscillate perpendicular to direction of energy transfer.
   (b) Progressive: point of maximum displacement moves in direction of transfer of energy.
   Stationary: points of maximum and minimum amplitude are fixed.
   (c) Node: point of zero amplitude in a stationary wave.
   Antinode: point of maximum amplitude in a stationary wave.
3. UHF radio, microwaves, visible light, ultraviolet, gamma rays.
4. The radiation is in the form of a transverse wave. Unpolarised transverse waves contain vibrations in all planes perpendicular to the direction of propagation of energy. The process of polarisation selects one of these planes. Longitudinal waves contain vibrations in the direction of energy so no further selection of oscillations is possible.
5. $0.9\,\text{W m}^{-2}$ (using inverse square law)
6. (a) See 2.4.3 Interference.
   (b) Two sources must be coherent (same frequency and constant phase difference) and have similar amplitudes to produce good contrast between interference maxima and minima.
7. Diffraction spreads light form a single slit so that it illuminates two slits to produce coherent sources. Diffraction spreads out the light from two slits so that the beams overlap and produce interference patterns.
8. 1.1 mm
9. (a) At the maximum position the waves are arriving in phase and interfering constructively. By moving 1.8 m the wave from the lower speaker travels half a wavelength further than the other so that the waves are antiphase and interfere destructively.
   (b) 1.6 m
10. 7.3°

11. **(a)**

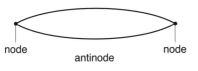

        node   node

               antinode

   (b) 600 Hz; 0.43 m

12. 342 m s^{-1}

Unit 2 Module 5

1.   **(i)** 3.0×10^{-19} J

     (ii) 1.0×10^{15} Hz

2. 2.9×10^{15} Hz

3. 1.88×10^{-20} J

4. 5.7×10^{7} m

5. 1.33×10^{-15} J

Data, formulae and relationships

Data

Values are given to three significant figures, except where more are useful.

speed of light in a vacuum	c	3.00×10^8 m s^{-1}
permittivity of free space	ε_0	8.85×10^{-12} C^2 N^{-1} m^{-2} (F m^{-1})
elementary charge	e	1.60×10^{-19} C
Planck constant	h	6.63×10^{-34} J s
gravitational constant	G	6.67×10^{-11} N m^2 kg^{-2}
Avogadro constant	N_A	6.02×10^{23} mol^{-1}
molar gas constant	R	8.31 J mol^{-1} K^{-1}
Boltzmann constant	k	1.38×10^{-23} J K^{-1}
electron rest mass	m_e	9.11×10^{-31} kg
proton rest mass	m_p	1.673×10^{-27} kg
neutron rest mass	m_n	1.675×10^{-27} kg
alpha particle rest mass	m_α	6.646×10^{-27} kg
acceleration of free fall	g	9.81 m s^{-2}

Conversion factors

unified atomic mass unit	1 u = 1.661×10^{-27} kg
electron-volt	1 eV = 1.60×10^{-19} J
	1 day = 8.64×10^4 s
	1 year $\approx 3.16 \times 10^7$ s
	1 light year $\approx 9.5 \times 10^{15}$ m

Mathematical equations

arc length $= r\theta$

circumference of circle $= 2\pi r$

area of circle $= \pi r^2$

curved surface area of cylinder $= 2\pi rh$

volume of cylinder $= \pi r^2 h$

surface area of sphere $= 4\pi r^2$

volume of sphere $= \dfrac{4}{3}$

Pythagoras' theorem: $a^2 = b^2 + c^2$

For small angle $\theta \Rightarrow \sin \theta \approx \tan \theta \approx \theta$ and $\cos \theta \approx 1$

$\lg(AB) = \lg(A) + \lg(B)$

$\lg\left(\dfrac{A}{B}\right) = \lg(A) - \lg(B)$

$\ln(x^n) = n \ln(x)$
$\ln(e^{kx}) = kx$

Formulae and relationships

Unit 1 – Mechanics	**Unit 2 – Electrons, waves and photons**
$F_x = F \cos\theta$ $F_y = F \sin\theta$	$\Delta Q = I\Delta t$
$a = \dfrac{\Delta v}{\Delta t}$	$I = Anev$
$v = u + at$	$W = VQ$
$s = \dfrac{1}{2}(u + v)t$	$V = IR$
$s = ut + \dfrac{1}{2}at^2$	$R = \dfrac{\rho L}{A}$
$v^2 = u^2 + 2as$	$P = VI \quad P = I^2R \quad P = \dfrac{V^2}{R}$
$F = ma$	$W = VIt$
$W = mg$	e.m.f. $= V + Ir$

moment $= Fx$

$$V_{out} = \frac{R_2}{R_1 + R_2} \times V_{in}$$

torque $= Fd$

$v = f\lambda$

$$\rho = \frac{m}{V}$$

$$\lambda = \frac{ax}{D}$$

$$p = \frac{F}{A}$$

$d \sin\theta = n\lambda$

$W = Fx \cos\theta$

$$E = hf \quad E = \frac{hc}{\lambda}$$

$$E_k = \frac{1}{2} mv^2$$

$hf = \phi + KE_{max}$

$E_p = mgh$

$$\lambda = \frac{h}{mv}$$

efficiency $= \dfrac{\text{useful energy output}}{\text{total energy input}} \times 100\%$

$R = R_1 + R_2 + \ldots$

$F = kx$

$$\frac{1}{R} = \frac{1}{R_1} = \frac{1}{R_2} + \ldots$$

$$E = \frac{1}{2} Fx \quad E = \frac{1}{2} kx^2$$

$$\text{stress} = \frac{F}{A}$$

$$\text{strain} = \frac{x}{L}$$

$$\text{Young modulus} = \frac{\text{stress}}{\text{strain}}$$

Index